CENTRAL INTELLIGENCE SYSTEMS OF ATOM

PARTICLE CLOUD CIRCULATION SYSTEMS INSIDE ELECTRONS, NEUTRONS, AND PROTONS

EZZAT E. MAJD POUR, M.D.

outskirts press

Central Intelligence Systems of Atom
Particle Cloud Circulation Systems Inside Electrons, Neutrons, and Protons

V2.0

Outskirts Press, Inc.
http://www.outskirtspress.com

ISBN: 978-1-9772-3771-2

PRINTED IN THE UNITED STATES OF AMERICA

A NOTE FROM AUTHOR

This book is the results of personal research works and new discoveries of the author. Which no-body has discovered these findings during the past world history, except the author who has done those during the last several decades, through continuous research works.

EZZAT E. MAJD POUR, M.D.

INDEX

PAGES

DEFINITION OF PARTICLE CLOUDS

THE FUNDAMENTAL PARTICLES HAVE TENDENCIES TO CONSTRUCT PARTICLE - CLOUDS COPIES FROM ENVIRONMENTAL SUBJECTS, IN WHICH THE PRODUCED PARTICLE - CLOUD COPIES ARE EXACTLY SIMILAR AND THE SAME DUPLICATE PARTICLE –MADE COPY FROM ORIGINAL SUBJECT, BUT MADE BY PARTICLES.

THESE PARTICLE CLOUD –MADE COPIES FROM DIFFERENT SUBJECTS POSSESSES ALL INFORMATION AND IMAGE COPIES EXACTLY THE SAME AS ORIGINAL SUBJECTS, THESE CLOUDS CALLED THE FUNDAMENTAL PARTICLES INFORMATION IMAGE PARTICLE CLOUDS (FP – I.I. – P. Cl.).
THE EMITTED FP – I.I. – P. Cl. FROM SUBJECTS REVEALS EXACTLY THE SAME, AND ALL BEHAVIORS, MOVEMENTS, ACTS, PSYCHOLOGICAL, PHYSIOLOGICAL, BIOLOGICAL REPRESENTATIONS, KNOWLEDGES, ETHICAL, SOCIAL, RELIGIOUS, SCIENCE, INFORMATIONS AND IMAGES, TEACHING AND TRAINING CAPABILITIES, AS WELL AS LEARNING ABILITIES.
POSSESSION OF ABOVE PROPERTIES BY FP – I.I. –P. Cl MAKE THEM THE MOST IMPORTANT TOOLS IN TEACHING, TRAINING, AND EDUCATIONS IN HOME, SOCIETY, POLITICS, SCHOOLS, FROM PRIMARY TO HIGH SCHOOL, AND FROM COLLEGE TO UNIVERSITIES AND POST GRADUATE EDUCATIONS ALL OTHERS EDUCATIONAL SYSTEMS.

THE PROVIDER SUBJECT'S FP – I.I. – P. Cl. EXHIBITS THE STATUS OF SUBJECT'S MENTALITIES, EITHER NORMALITY OF SUBJECTS PSYCHE, OR ABNORMALITIES OF THE PROVIDER SUBJECTS PSYCHEAS, WHICH THE PROVIDER SUBJECTS ARE TRANSMITTING POSITIVE NORMAL PRODUCTIVE FP – I.I. – P. Cl. OR THEY ARE TRANSMITTING ABNORMAL PSYCHIATRIC DISEASES FROM SICK TO NORMAL, THROUGH THE TRANSMISSION OF THE FP – I.I. – P. Cl.

THESE ARE THE FP – I.I. – P. Cl. CHRACTORS, WHEN THE SUBJECT MOVES, ACTS, SPEAKS, TEACH, LISTEN, ORDER, FIGHT, OR STAY SILENT, SIMULTANEOUSLY THE FP – I.I. – P. Cl. STARTS TO DO EXACTLY THE SAME PERFORMENCES OF THE ORIGINAL SUBJECTS, EXACTLY SAME THE FP – I.I. – P. Cl. ALSO START TO MOVE, TO SPEAK, TO ACT, FIGHT, OR STAY SILENT, EXACTLY THE SAME AS ORIGINAL SUBJECT, THE FUNDAMENTAL PARTICLES INFORMATION – IMAGE PARTICLE CLOUDS ARE SUBJECT – CLOUDS.
THE FP – I.I. – P. Cl. BUILD PSYCHE OF THE ELECTRONS, NUCLEONS, AND ATOMS OF THE CNS ATOMS, THROUGH STORING THE EXTERNAL ENVIRONMENTAL KNOWLEDGE, INFORMATIONS, IMAGES INSIDE THE CNS NANO – UNIT CONSTRUCTIONS OF THE RECIPIENTS.
IN GENERAL, IN MOST CASES, THERE ARE SIMILARITIES EXIST BETWEEN THE PROVIDER SUBJECTS INFORMATION IMAGE PARTICLE CLOUDS (EITHER IF THESE PARTICLE CLOUDS ARE NORMAL OR ABNORMAL) AND RECIPIENT SUBJECT'S PRODUCED CNS INTER ELECTRON NUCLEON FP – I.I. – P. Cl. COMPOUND CONSTRUCTIONS,
THIS PHENOMENON IS PSYCHE GENESIS OF THE CNS ELECTRONS, NUCLEONS, ATOMS, AND NANO – UNITS. THE THOUGH CURRENTS, AND SOUL (PSYCHE) OF THE ANIMAL AND LIVING THING SPECIES ALL ARE INTERNAL CNS ATOMS FP – I.I. – P. Cl. CIRCULATION SYSTEMS AND INTERACTIONS WITH EACH OTHER, UNDER A.S. I. FP. Mol. C.I.C.
WHEN THE ENTIRE TOTAL ELECTRONS NUCLEON POPULATIONS OF THE CNS INTERACT WITH EACH OTHER THROUGH A. S. I. FP. Mol. C.I.C. BETWEEN DIFFERENT CENTERS OF CNS CENTERS, THESE PRTICLE CLOUD INTERACTIONS WITH EACH OTHER AS WELL AS WITH CNS DIFFERENT CENTERS ELECTRONS AND NUCLEONS WITH EACH OTHER, BIOLOGICALLY ALL OF THESE INTERACTIONS AND PARTICLE CLOUD CURRENTS SENSE, AS THOUGHT CURRENTS AND THOUGHT CURRENT SYSTEMS OR PSYCHE BY LIVING THINGS INSIDE THEIR HEADS.

ELECTRONS, NUCLEONS, ATOMS, AND CELLS

FP – I.I. – P. Cl. transmissions are contagious and can cause particle cloud infection at recipients

THE PHENOMENON OF REPLICATION OF THE ELECTRON CONSTRUCTIONS, AND REPLICA NUCLEON GENESIS

THE CAUSES OF THE REPLICA PSYCHE GENESIS IN LIVING THINGS

Exogenous particle circulation system (EX. PCS)

THE FP – I.I. – P. Cl. ARE CONTAGIOUS, PARTICLE CLOUDS AFTER EMISSION FROM PROVIDER SUBJECTS TRAVEL AIRBORNE, AND EXCHANGE PARTICLE CLOUDS BACK AND FORTH, BETWEEN PARTICLE CLOUD PROVIDER SUBJECTS CNS ATOMS, AND PARTICLE CLOUD RECIPIENT SUBJECTS CNS ATOMS, THE PARTICLE CLOUD EXCHAGE BETWEEN DIFFERENT SUBJECTS ATOMS IN EARTH, MOSTLY TAKE PLACE THROUGH AIRBORNE EXOGENOUS PARTICLE CLOUD CIRCULATIONS SYSTEMS (EX. PCS) PATHS, AND PARTICLE CIRCULATION ROUTES.
THE EXOGENOUS FUNDAMENTAL PARTICLE CIRCULATIONS SYSTEMS (EX. PCS) ARE THE MOST COMMON FORMS OF PARTICLE CURRENT TRANSMISSIONS ROUTES, WHICH USED BETWEEN DIFFERENT LIVING THINGS CNS ATOMS, TO EXCHANGE PARTICLE CLOUDS WITH EACH OTHER, AND COMMUNICATE WITH EACH OTHER IN DAILY BASES.

INDIGENOUS HIERARCHY SYSTEMS HIERARCHY SYSTEMS IN CELL'S, ATOMS, ELECTRONS, NUCLEONS, NANO UNITS, AND F.P.

THE HIERARCHY SYSTEMS IN BODY ORGANS AND SYSTEMS, HIERARCHY SYSTEMS IN LIVING THINGS

The Indigenous Particle Circulation Systems (I – PCS)

======================

INDIGENOUS PARTICLE CIRCULATION SYSTEMS (I-PCS) ARE PARTICLE CURRENT CIRCULATION SYSTEMS, THAT CIRCULATE BETWEEN LIVING THINGS TOTAL BODY ELECTRONS AND NUCLEONS IN ALL LIVING THING SPECIES, I – PCS CONNECTS THE TOTAL BODY ELECTRONS, NUCLEONS POPULATIONS TO EACH OTHER, THROUGH I – PCS TOTAL BODY ELECTRONS, NUCLEONS, AND ATOM POPULATIONS COMMUNICATE WITH EACH OTHER UNDER HIERARCHY FUNDAMENTAL PARTICLES SYSTEMS (F.P. - H), AS WELL AS UNDER HIERARCHY NANO – UNITS ORDERS SYSTEMS (N.U. - H), AND HIERARCHY ELECTRON, NUCLEON, ATOM ORDER - SYSTEMS (E.N.A. - H).

THE CELL'S HIERARCHY OPERATION ORDERS SYSTEMS (C – H) ALSO CONNECTED TO NANO – UNIT'S HIERARCHY ORDERS SYSTEM CENTERS AND ELECTRONS, NUCLEONS AND ATOMS HIERARCHY OPERATIONS SYSTEM CENTERS.
ALL OF THE EX. H.S., I. HS., N.U.- HS, AND C-H.S. WHICH ALL COLLECTIVELY CALLED INDIGENOUS HIERARCHY OPERATION SYSTEMS CENTERS (I – H), ALL TOGETHER FUNCTIONS WITH FULL COOPERATIONS AND COORDINATIONS OF EACH OTHER, UNDER FUNDAMENTAL PARTICLE'S PRECISIONS ACCURACIES.

Storage of S. Y. E. T. – FP – I.I. – P. Cl. inside CNS Electrons, Nucleons, Atoms

FACTUAL INTELLIGENCE (RI) GENESIS

Psyche genesis inside CNS Electrons,

Atom's Psyche genesis, Nucleon's Psyche genesis Genesis of intelligence in Electrons, Nucleons

THE RECIPIENT SUBJECT'S DIFFERENT SENSARY ORGANS ATOMS, THROUGH THE POWERFUL GRAVITON FORCES OF THEIR ATOM'S ELECTRONS, NUCLEONS TOWARD NEEDED CLOUDS AND PARTICLES, THEY ATTRACT SELECTED NEEDED EXOGENOUS FP –I.I. – P. Cl. FROM THEIR SURROUNDING AIR, TRANSMIT AND CARRY CAPTURED PARTICLE CLOUDS, TO THE RESPECTED NEEDED CNS DIFFERENT ATOM CENTERS, WHICH THESE CNS CENTER'S ATOMS, ELECTRONS, NUCLEONS ARE IN EXTREME NEEDS OF THESE EXOGENOUS PARTICLE CLOUDS, IN ORDER THESE ELECTRONS AND NUCLEONS CAN CONSTRUCT THEIR INTERNAL ELECTRON NUCLEON PARTICLE COMPOUND CONSTRUCTIONS, THROUGH THE USE OF THOSE NEEDED PARTICLES, AND INFORMATION IMAGE PARTICLE CLOUDS MOLECULES INSIDE THEIR ELECTRONS NUCLEON PARTICLE COMPOUND CONSTRUCTIONS. (ELECTRON GENESIS, NUCLEON GENESIS).

THE VISION CENTER CNS ELECTRONS, NUCLEONS, ATOMS, ARE IN NEEDS OF LIGHT FUNDAMENTAL PARTICLES, AND LIGHT FUNDAMENTAL PARTICLES INFORMATION AND IMAGES PARTICLE CLOUDS (Y. F.P. – I.I. – P. Cl.),
THE LIGHT PARTICLE SENSARY ORGANS ATOMS, FOLLOWING ATTRACTION, AND CAPTURE OF THE LIGHT PARTICLE CLOUDS, TRANSMIT THE Y. FP. – I.I. – P. Cl. INTO VISION CNS CENTERS INTERNAL ELECTRONS, NUCLEONS SUBSYSTEM UNITS CHEMICAL LAB.S, THERE THROUGH THE REGENERATIVE AUTONOMOUS SEQUENTIAL INTER FUNDAMENTAL PARTICLE MOLECULAR CHEMICAL INTERACTIONS CYCLES AND CHAINS (A. S. I. FP. Mol. C.I.C.) INSIDE THE VISION CENTER ELECTRONS, NUCLEONS SUBSYSTEM UNITS CHEMICAL LAB.S, THE INCOMING EXOGENOUS Y. FP – I.I. – P. Cl. COMBINE WITH CNS VISION CENTERS INTERNAL ELECTRON, NUCLEON PARTICLE COMPOUNDS, PRODUCE Y. FP. – I.I. – P. Cl. – COMPOUNDS, THROUGH THE USE OF THESE LIGHT PARTICLE COMPOUNDS CONSTRUCT VISION CENTER ELECTRONS, NUCLEON LIGHT PARTICLE CLOUD COMPOUND CONSTRUCTIONS, THESE ARE PHENOMENONS OF NEO – ELECTRONS GENESIS, AND NEO – NUCLEON GENESIS CONSTRUCTION PHENOMENONS OF THE VISION CNS CENTERS.

THE STORAGE OF Y. FP. – I.I. – P. Cl. COMPOUNDS INSIDE VISION CENTER ELECTRONS AND NUCLEONS, AS AN INFORMATION IMAGE REFERENCES ENCYCLOPEDIAS, THAT STORED INSIDE THE VISION CENTERS ELECTRONS AND NUCLEONS AS REFERENCES FOR CNS INTELLIGENCE SYSTEMS CENTERS, FOR THEIR FUTUTRE USE, AS THE VALID REFERENCES TO BE USED FOR DECISION MAKING , COMPARISONS, ORDERS, AND ACTS PROPERLY, WHEN THERE ARE FUTURE ENVIRONMENTAL EVENTS, AND INCIDENTS AT THE OUTSIDE WORLD, WHEN CNS COMMANDING SYSTEM CENTERS ELECTRONS, NUCLEONS NEEDS, AVAILABLE VALID REFERENCES INFORMATIONS AND IMAGES TO DECIDE AND ORDER. WHICH THESE PARTICLE CLOUDS ARE THERE TO BE USED BY CNS COMMANDERS AVAILABLE INSIDE THE ELECTRONS AND NUCLEONS FOR RETRIEVALS, REVIEWS AND DECISION MAKINGS.

IT IS EXACTLY THE SAME, IN REGARD TO THE AUDITARY CNS ATOM'S, ELECTRONS, NUCLEON'S NEEDS FOR SONIC FUNDAMENTAL PARTICLES, AND S. FP. – I.I. – P. Cl., IN ORDER TO USE THOSE SONIC PARTICLES, AND S. FP.- I.I.-P. Cl. IN THE CONSTRUCTION OF AUDITARY CNS ATOM'S INTER ELECTRONS, NUCLEONS SONIC PARTICLE CLOUD COMPOUND CONSTRUCTIONS WITH SONIC PARTICLE CLOUDS. AND CREATE NEO – ELECTRONS GENESIS AND NEO – NUCLEON GENESIS THROUGH THE USE OF THE EXOGENOUS S. FP – I.I. – P. Cl. COMPOUND CONSTRUCTIONS. THE STORAGES OF THE S. FP. – I.I. – P. Cl. – COMPOUNDS (STORAGE) SYSTEMS ARE NEEDED AS REFERENCE ENCYCLOPEDIAS FOR FUTURE USES BY CNS DIFFERENT CENTER SYSTEMS.
ABOVE PATTERNS ALSO APPLY FOR ALL OTHER BIO FRIENDLY FUNDAMENTAL PARTICLE CLOUDS, SUCH AS THERMAL PARTILES CLOUDS, ELECTRIC PARTICLES CLOUDS, ETC. THE RULES FOR ALL ARE THE SAME AS ABOVE.

THESE ARE PHENOMENONS OF STORAGE OF THE EXOGENOUS ENVIRONMENTAL INFORMATION IMAGE PARTICLE CLOUDS IN MOLECULAR CONSTRUCTION FORMS INSIDE THE CNS ELECTRONS, NUCLEONS,
THESE ARE ALSO PHENOMENON OF PSYCHE GENESIS OF ELECTRONS, AND NUCLEONS PSYCHE GENESIS PHENOMENONS, IN CNS ATOMS, AND PRODUCTIONS OF THE INTELLIGNCE ELECTRONS, NUCLEONS AND ATOMS.

REPLICATIONS OF THE CONSTRUCTIONS OF THE ELECTRONS, NUCLEONS, AND ATOMS

Replica Electron genesis and Replica Nucleon genesis inside CNS Atoms

TRANSMISSION OF REPLICA BEHAVIORS FROM PARENTS AND TEACHERS TO THE CHILDREN AND STUDENTS

TWIN ELECTRON GENESIS AND TWIN NUCLEON GENESIS

REPLICA – PSYCHE GENESIS

REPLICATIONS OF THE PROVIDER'S SUBJECT'S PSYCHE COPY, INSIDE THE RECIPIENT SUBJECT'S BEHAVIORS

THE PARENTS, TRAINERS, AND TEACHERS CONTINUOUSLY, IN DAILY BASIS, EMIT AND TRANSMIT THEIR INTERNAL CNS PRODUCED Y. S. T. E. - FP. – I.I. – P. Cl. TO THEIR RECIPIENT CHILDRENS, STUDENTS MONTHS, AND YEARS.

THE PARENT'S AND TEACHER'S Y. S. E. T. – FP. - I.I. – P. Cl. IN DAILY BASIS, TRAVEL AIRBORNE, AND INTER INTO CHILDREN'S AND STUDENT'S CNS ELECTRONS, NUCLEONS SUBSYSTEM UNITS CHEMICAL LAB.S, AND UNDER REGENERATIVE A. S. I. FP. MOL. C.I.C. THE PARENTS AND TEACHERS PARTICLE CLOUDS COMBINES WITH CHILDRENS AND STUDENTS INTER ELECTRON, NUCLEON PARTICLE COMPOUNDS MOLECULES, AND CONSTRUCT THE CHILDREN'S AND STUDENT'S CNS ELECTRONS AND NUCLEONS S. Y. E. T. - FP. - I.I. - P. Cl. - COMPOUND CONSTRUCTIONS.

THE NEWLY CONSTRUCTED S. Y. E. T. – FP – I.I. – P. Cl. - COMPOUNDS CONSTRUCTION'S PARTICLE CLOUDS MOLECULES, INSIDE THE CHILDRENS AND STUDENTS CNS ELECTRONS AND NUCLEONS MOSTLY ARE CONSTRUCTED, AND EMITTED PARTICLE CLOUDS FROM PARENTS, AND TEACHERS CNS ELECTRONS AND NUCLEONS PARTICLE CLOUD COMPOUND CONSTRUCTIONS, AND BOTH HAVE SIMILARITY.
THE S. Y. E. T. – FP – I.I. – P. Cl. – COMP. COMSTRUCT THE ELECTRONS AND NUCLEONS PARTICLE CLOUD MOLECULAR COMPOUND CONSTRUCTIONS, THEREFORE THE CONSTRUCTED ELECTRONS, NUCLEONS IN CHILDREN'S CNS POSSESSES SIMILARITIES TO THE PARENTS AND TEACHERS ELECRONS, NUCLEONS.

THE CREATED ELECTRONS, NUCLEON PARTICLE COMPOUND CONSTRUCTIONS OF THE CHILDREN AND PARENT, OR STUNDENTS AND TEACHER BOTH ARE SIMILAR - PATICLE CLOUD INFORMATION IMAGE CONSTRUCTED MOLCULAR STRUCTURES, THE PARTICLE CLOUD PROVIDERS AND RECIPIENT INDIVIDUALS CHARACTORS, KNOWLEDGE, BEHAVIORS, THE ACTS, THE SCIENCE AND KNOWLEDGE ENCYCLOPEDIA CONTENTS, ETC. ALL HAVE SIMILARITIES, BECAUSE THEY HAVE SIMILAR CNS ELECTRONS, NUCLEONS FP. – I.I. – P. Cl. COMPOUND CONSTRUCTIONS.

THIS IS THE PHENOMENON OF THE INTELLIGENCE ELECTRON – NUCLEON GENESIS, AND INTELLIGENCE ATOM GENESIS PHENOMENON. IT IS ALSO, THE STORAGE OF THE OUTSIDE ENVIRONMENTAL PARTICLE CLOUDS INFORMATION AND IMAGES, AT INSIDE THE CNS ATOMS IN PARTICLE CLOUD COMPOUND FORMS INSIDE THE RECIPIENTS CNS ELECTRONS, NUCLEONS.

THE REPLICATE PSYCHE AND BEHAVIOR GENESIS

THE TWIN PARTICLE CLOUD COMPOUND GENESIS IN PROVIDERS AND RECIPIENTS CNS ELECTRONS, NUCLEONS

TWIN ELECTRON, NUCLEON GENESIS

THE PARENTS AND TEACHERS, THROUGH PERSISTENT PROLONGED MONTHLY, YEARLY AIRBORN S. Y. T. E. - FP – I.I. – P. Cl. EMISSIONS FROM THEMSELVES, AND CONTINUOUS TRANSMISSION OF THESE PRODUCED AIRBORN INFORMATIONS AND IMAGES PARTICLE CLOUDS TO THE RECIPIENT CHILDREN AND STUDENT'S INTERNAL CNS ELECTRONS, NUCLEONS SUBSYSTEM UNITS CHEMICAL LAB.S, CAUSE COMBINING OF THE PARENT'S AND TEACHER'S PRODUCED AND EMITTED EXOGENOUS PARTICLE CLOUDS, WITH CHILDRENS

AND STUDENTS INTERNAL CNS ELECTRONS, NUCLEONS PARTICLE COMPOUNDS IN DAILY BASIS. AND THIS PROCESS CONTINUOUSLY CONTAMINATE THE CHILDRENS AND STUDENTS BRAINS ELECTRONS AND NUCLEONS CONSTRUCTIONS.

PARENT AND TEACHERS THROUGH PROLONGED EMITTED AND TRANSMITTED FUNDAMENTAL PARTICLE INFORMATION IMAGE PARTICLE CLOUDS, STORAGE AND DEPICTIONS OF THOSE PARTICLE CLOUDS INSIDE THE STUDENTS CNS ELECTRONS AND NUCLEONS IN PARTICLE CLOUD COMPOUND STORAGE FORMS. CAUSE THE PRODUCTIONS OF THE RECIPIENT SUBJECTS PSYCHE, AND THOUGHT CURRENTS PROCESS (STORED S. Y. T. E. – P – I.I. – P. Cl. – COMP. CONSTRUCTIONS OF THE ELECTRONS AND NUCLEONS), TO BE ACCORDING AND SIMILAR TO PROVIDER SUBJECTS FUNDAMENTAL PARTICLE CLOUD TRANSMITTED ORDERS, AND PARTICLE CLOUD COMPOUND CONSTRUCTIONS OF RECIPIENTS, ARE SIMILAR TO PROVIDERS SUBJECTS.

PARENTS AND TEACHERS EMIT PARTICLE CLOUD FROM THEMSELVES IN DAILY BASIS, THE S. Y. E. T. – FP – I.I. – P. Cl. TRAVEL AIRBORN FROM PARENTS, TEACHERS TO CHILDREN AND STUDENTS, INTER INTO THE CHILDREN'S STUDENTS INTER CNS ELECTRONS NUCLEONS SUBSYSTEM CHEMICAL LAB.S, COMBINES WITH INTERNAL CNS ELECTRONS, NUCLEONS PARTICLE COMPOUNDS, PRODUCE S. Y. E. T. – FP – I.I. – P. Cl. COMPOUND CONSTRUCTIONS OF THE RECIPIENTS CNS ELECTRONS AND NUCLEON, IDENTICAL OR SIMILAR TO PROVIDERS.
UNDER PARENTS AND TEACHERS PERSISTENT PROGRESSIVE DAILY CONTINUOUS S. Y. T. E. – FP – I.I. – P. Cl. TRANSMISSIONS, THROUGH INTERNAL CNS ELECTRONS NUCLEONS PARTICLE CLOUD COMPOUND CONSTRUCTIONS, THE RECIPIENTS CONSTRUCT THE REPLICA CNS ELECTRONS, NUCLEONS, AND ATOM'S PARTICLE CLOUD COMPOUND CONSTRUCTIONS, INSIDE THEIR CNS ELECTRONS AND NUCLEONS. WHICH ARE MOSTLY SIMILAR AND IDENTICAL PARTICLE CLOUD COMPOUNDS CONSTRUCTIONS, NEO ELECTRONS – NUCLEONS CONSTRUCTIONS SIMILAR TO PROVIDER SUBJECTS.

TWIN ELECTRONS NUCLEON GENESIS,

Identical Psyche Genesis,

Similar Thought Current Genesis

NORMAL PSYCHE PARENTS PRODUCE NORMAL KIDS, ABNORMAL PARENTS PRODUCE ABNORMAL CHILDREN

THE TEACHERS AND PARENTS GRADUALLY CONSTRUCT THE STUDENTS AND CHILDRENS CNS NEW ELECTRONS NUCLEONS PARTICLE CLOUD COMPOND CONSTRUCTIONS, SIMILAR TO THE PARTICLE CLOUD COMPOUNDS CONSTRUCTIONS OF THEMSELVES. BOTH WITH SIMILAR NEW IDENTICAL PARTICLE CLOUD COMPOUNDS NEW PSYCHE GENESIS, AND THOUGHT CURRENT GENESIS. THE INDENTICAL MOLECULAR S. Y. T. E. – FP – I.I. – P. Cl. – COMPOUND CONSTRUCTIONS NEO THOUGHT CURRENT SYSTEMS CONSTRUCTIONS.
UNDER THESE CASES, THE ELECTRONS AND NUCLEONS AND ATOMS CONSTRUCTIONS IN RECIPIENT SUBJECTS CNS ELECTRONS, NUCLEONS, IS TWIN OR SIMILAR TO THE PROVIDER SUBJECT'S CNS ELECTRONS AND NUCLEON'S PARTICLE CLOUD COMPOUND CONSTRUCTIONS, WHICH AS THE RESULTS PRODUCE AND CREATE A SIMILAR TYPES TWIN PSYCHE- GENESIS, AND IDENTIAL TWIN THOUGHT CURRENT GENESIS IN BOTH.

UNDER THIS PHENOMENON, THE RECIPIENT'S PARTICLE CLOUD COMPOUND CONSTRUCTIONS, AT CNS INTERNAL ELECTRON, AND NUCLEON'S CONSTRUCTIONS, THE CONSEQUENTIAL PSYCHE GENESIS, BEHAVIOR GENESIS, THOUGHT CURRENT GENESIS ALL RELATIVELY ARE SIMILAR, IDENTICAL, SIMILAR TO THE ELECTRONS, AND NUCLEON'S PARTICLE CLOUD COMPOUNDS CONSTRUCTIONS. THE ABOVE PHENOMENON IS BEHAVIORS GENESIS, PSYCHE GENESIS OF RECIPIENT'S PSYCHE, WITH PROVIDER SUBJECT'S PARTICLES PARTICLE CLOUDS, AND AS RESULTS CAUSE IDENTICAL BEHAVIOR GENESIS. THESE SIMILARITIES IN NATURAL PHYSICAL, CHEMICAL, BIOLOGICAL OCCURRENCES, MOSTLY ARE RELATIVE SIMILARITIES, AND ARE NOT MATHEMATICAL ONES.

UNDER THESE "PROVIDER – RECIPIENTS" CNS INTERNAL ELECTRON NUCLEON PARTICLE CLOUD COMPOUNDS PRODUCTIONS, IN FINAL RESULTS THE SIMILAR CNS PARTICLE CLOUD MOLECULAR CONSTRUCTIONS OF BOTH GROUPS, CREATE AND CAUSES SIMILARIES AT THOUGHT CURRENT GENESIS, AND SIMILARITIES AT PSYCHE PRODUCTIONS IN BOTH RECIPIENTS GROUPS AND PROVIDER GROUPS.

UNDER THESE PHENOMENONS, THE PSYCHE – CONSTRUCTIONS OF THE PROVIDER SUBJECTS PSYCHES, ALL LOOKS SIMILAR TO THE RECIPIENT SUBJECTS PSYCHE AND THOUGHT CURRENTS. THAT IS THE PARENTS AND TEACHER'S PARTICLE CLOUDS, THAT CONSTRUCT THE CHILDRENS AND STUNDENTS PSYCHE CONTENTS, WHICH CAUSE BOTH POSSESSES SIMILARITIES PSYCHE AND PARTICLE CLOUD CONSTRUCTIONS.

THIS IS THE REASON, THE NORMAL PARENTS AND TEACHERS PRODUCE NORMAL PSYCHE, AND NORMAL THOUGHT CURRENTS CHILDREN. IN CONTRARY THE MENTALLY ABNORMAL PARENTS AND TEACHERS PRODUCE MENTALLY DISORDERED AND ABNORMAL PSYCHE CHILDREN AND STUDENTS, AND THESE CHILDRENS AND STUDENTS PSYCHE AND THOUGHT CURRENTS AS WELL AS THEIR INTERNAL CNS PARTICLE CLOUD MOLECULAR CONSTRUCTIONS POSSESSES SIMILARITIES WITH EACH OTHERS.
THIS PHENOMENON CAUSE BEHAVIORS, PSYCHES, AND THOUGHT CURRENTS OF THE PROVIDERS AND RECIPIENTS TO DEVELOP SIMILAR TO EACH OTHER, BOTH POSSESSES THOUGHT CURRNTS GENESIS, AND PSYCHE GENESIS SIMILARITIES.

Rejections of Identical Psyche Genesis

Immunity Systems, and Defensive Systems Against Particle Cloud Infections

THE PARENTS, AND TRAINER'S ATTEMPTS TO BUILD NEW REPLICA PARTICLE CLOUD COMPOUNDS CONSTRUCTIONS INSIDE THE CHILDREN'S AND STUDENTS INTERNAL CNS ELECTRONS, NUCLEONS, AND CONSEQUENTIAL NEW REPLICATED PSYCHE GENESIS, AND THOUGHT CURRENT GENESIS IN CHILDREN'S AND STUDENTS BRAINS, SIMILAR TO PARENTS AND TEACHERS, SOME TIMES THIS PROCESS OF LEARNING FAILS, SECONDARY TO STUDENTS AND RECIPIENT SUBJECTS PREVIOUSLY CONSTRUCTED DEFENSIVE AND IMMUNITY SYSTEMS CONSTRUCTIONS WHICH CONSTRUCTED INSIDE ELECTRONS NUCLEONS IN PARTICLE CLOUD COMPOUND FORMS FROM BEFORE.
THIS PHENOMENON HAS THE UTMOST IMPORTANCES IN TREATMENTS, AS WELL AS IN PREVENTIONS AGAINST WRONG INFECTED CONTAMINANT PARTICLE CLOUDS COMPOUNDS CONSTRUCTIONS INSIDE RECIPIENTS BRAINS.
IN MOST ORDINARY AND NORMAL CONDITIONS, THE CHILDREN'S CNS INTERNAL ELECTRONS, NUCLEONS PARTICLE CLOUD COMPOUND CONSTRUCTIONS ARE REPLICA OF THE PARENTS AND TEACHERS EMITTED PARTICLE CLOUDS, AND THE PSYCHE GENESIS IN RECIPIENTS ARE SIMILAR TO PROVIDERS, THIS IS TWIN PSYCHE GENESIS PHENOMENON BETWEEN PARENTS AND CHILDREN.

UNDER TWIN PARTICLE CLOUDS COMPOUNDS, AND TWIN PSYCHE GENESIS PHENOMENON THE NORMAL PARENTS AND TRAINERS PRODUCE NORMAL PSYCHE CHILDRENS SIMILAR TO THEMSELVES MOSTLY, ALSO THE ABNORMAL, OR SICK, OR DEVIATED, OR CRIMINAL TRAINERS, TEACHERS, AND PARENTS ALSO TEACH, EDUCATE, AND CONSTRUCT ABNORMAL PARTICLE CLOUD CONSTRUCTIONS INSIDE THE RECIPIENTS CNS ELECTROS NUCLEONS, AND CREATE WRONG ABNORMAL PARTICLE CLOUD STORAGE INSIDE CNS ELECTRONS AND NUCLEONS OF STUDENT, SIMILAR TO TEACHERS AND EDUCATORS.

THE COMBINATIONS OF PARENTS AND TEACHERS EXOGENOUS FP – I.I. – P. Cl. WITH CHILDRENS AND STUDENT RECIPIENTS CNS INTERNAL ELECTRON NUCLEON PARTICLE COMPOUNDS, CAUSE THE RECIPIENT'S CNS ATOMS - CENTRAL INTELLIGENCE SYSTEM CENTERS (CIS), ELECTRONS CIS, AND NUCLEONS CIS CREATE THE CHILDREN'S PSYCHE AND THOUGHT CURRENTS, AND TRANSFORM THE RECIPIENTS PSYCHE SYSTEMS, AND THOUGHT CURRENT SYSTEMS, AND TRANSFORMATIONS ACCORDING TO THE EDUCATORS EMMITED PARTICLE CLOUDS PATTERNS AND ORDERS.
AND CHANGE ACCORDING THE PROVIDER SUBJECT'S INFORMATIONS AND IMAGES AND PSYCHE PATTERNS.

IN SOME CASES, BECAUSE OF THE CHILDREN'S AND STUDENTS PRE –EXISTING PARTICLE CLOUD COMPOUND CONSTRUCTIONS INSIDE THE INTERNAL ELECTRONS NUCLEONS, AND EXISTENCES OF STRONG CONSTRUCTED PRE –EXISTED PSYCHE AND THOUGHT PARTICLE CLOUD MOLECULAR CONSTRUCTIONS. WHICH ALL OF THESE ARE PREVIOUS, CONSTRUCTED AND DEVELOPED PSYCHE IMMUNITY AND DEFENSIVE SYSTEMS, UNDER THESE CONDITIONS THE RECIPIENT SUBJECTS PSYCHE ARE IMMUNE, AND CONTAMINATION PROOF AGAINST INCOMING MALICIOUS PARTICLE CLOUD INFLICTIONS. TRANSMISSIONS OF ABNORMAL PARTICLE CLOUDS, FROM ONE TO OTHER. MAY NOT TAKE PLACE, THIS PHENOMENON PLAYS MAJOR ROLES IN PREVENTIONS, AGAINST INFLICTION TO MENTAL DISEASES, WHICH SOME INDIVIDUALS ARE MORE PRONE THAN THE OTHERS, CAN ACT EFFECTIVELY AGAINST PARTICLE CLOUD CONTAMINANST. PREVIOUSLY DEVELOPED PARTICLE CLOUD IMMUNITY SYSTEMS, IS IMPORTANT FACTOR AGAINST PROVIDERS ABNOXIOUS INCOMING PARTICLE CLOUDS, NEUTRALIZE CONTAMINATION THROUGH PRE –EXISITNG PARTICLE INFORMATIONS AND IMAGES PARTICLE CLOUD COMPOUNDS CNS ELECTRONS NUCLEONS PREVIOUS MOLECULAR CONSTRUCTIONS, UNDER THE A.S. I. FP. Mol. C.I.C.

THE FP – I.I. – P. Cl. ARE CONTAGIOUS
THE NORMAL PARTICLE CLOUDS TRANSMISSION, PRODUCE NORMAL PSYCHE AND THOUGHT CURRENTS
THE ABNORMAL PARTICLE CLOUDS TRANSMISSION, PRODUCE ABNORMAL PSYCHE, AND PSYCHIATRIC DISEASES

Mental disorders are Abnormal S. Y. T. E. –FP – I.I. – P. Cl. transmitted diseases

THE PROVIDER SUBJECTS WITH ABNORMAL PARTICLE CLOUD COMPOUND CONSTRUCTION ELECTRONS NUCLEONS CNS ATOMS, CREATE AND EMIT AIRBORNE SONIC, LIGHT, THERMAL, ELECTRIC FUNDAMENTAL PARTICLE INFORMATION IMAGE PARTICLE CLOUDS (S. Y. E. T.– FP. - I.I. – P. Cl.) INTO ATMOSPHERE, THESE PARTICLE CLOUDS ARE FREE FLOATING MOLECULAR PARTICLE CLOUDS, COMPOSITE FUNDAMENTAL PARTICLES, AND PARTICLE CLOUDS INFORMATION IMAGE CLOUDS, TRAVELING FREELY IN SPACE IN DIFFERENT DIRECTIONS.

THE RECIPIENT SUBJECTS SENSARY ORGAN'S ELECTRONS NUCLEONS, THROUGH THEIR POWERFUL ATTRACTIVE GRAVITON FORCES, ATTRACT FROM ATMOSPHERE THE NEEDED PARTICLE CLASSES OF PROVIDER SUBJECTS FROM SPACES PARTICLE CLOUDS SUPPLIES, AND TRANSIT THE CAPTURED PARTICLE CLOUDS INTO CNS INTERNAL ELECTRONS NUCLEONS SUBSYSTEM UNITS CHEMICAL LAB.S. INSIDE RECIPIENTS CNS ELECTRONS NUCLEONS SUBSYSTEM UNITS CHEMICAL LAB.S, THROUGH THE REGENERATIVE A. S. I. FP. Mol. C.I.C. THE PROVIDER SUBJECTS INCOMING ABNORMAL EXOGENOUS S. Y. E. T. - FP – I.I. – P. Cl. COMBINE WITH RECIPIENTS INTER ELECTRON NUCLEONS PARTICLE COMPOUNDS, AND PRODUCE THE RECIPIENT SUBJECTS S. Y. T. E. - FP – I.I. – P. Cl. – COMPOUNDS, AND CONSTRUCT THE RECIPIENT SUBJECT'S ELECTRONS, NUCLEONS PARTICLE COMPOUND MOLECULAR CONSTRUCTIONS, WITH INCOMING EXOGENENOUS ABNORMAL PARTICLE CLOUDS, WHICH HAS BEEN PRODUCED BY MENTALLY SICK PARTICLE CLOUD PROVIDER PATIENS CNS ELECTRONS AND NUCLEONS.
THIS PHENOMENON IS STORAGE OF THE ABNORMAL FUNDAMENTAL PARTICLE CLOUDS INSIDE THE CNS INTERNAL ELECTRONS NUCLEONS MOLECULAR CONSTRUCTIONS. AND ABNORMAL FUNDAMENTAL PARTICLE CLOUD COMPOUNDS CONSTRUCTIONS IN RECIPIENTS.
THE AIR BORN TRANSMISSION OF THE ABNORMAL FP – I.I. – P. Cl FROM PROVIDER SICK SUBJECTS CNS ATOMS, TO THE RECIPIENT SUBJECTS CNS ELECTRONS, NUCLEONS, CAN CAUSE PARTICLE CLOUD INFECTIONS OF THE RECIPIENT'S ELECTRONS AND NUCLEONS. THESE TRANSMISSIONS CHANGES THE PARTICLE CLOUD COMPOUND CONSTRUCTIONS OF THE RECIPIENTS CNS ELECTRONS NUCLEONS CONSTRUCTIONS, TO BECOME SIMILAR TO THE PROVIDER PARENTS AND TEACHERS INTERNAL ATOM PARTICLE COMPOUND CONSTRUCTIONS.

THE ABNORMAL S. Y. E. T. – FP – I.I. – P. Cl. COMPOUND CONSTRUCTIONS OF CHILDREN'S AND STUDENT'S CNS INTERNAL ELECTRONS, NUCLEONS PARTICLE CLOUD COMPOUNDS. CAUSE THE CHILDRENS AND STUDENTS DEVELOPE NEW PSYCHE GENESIS, AND THOUGHT CURRENTS – GENESIS SIMILAR TO THE ABNORMAL PROVIDER PARENT'S PARTICLE CLOUD MOLECULAR NEW PSYCHES GENESIS AND NEO - THOUGHT CURRENTS GENESIS.
THAT IS THE REASON THE DEPRESSIVE PATIENTS TRANSMIT AND CREATE DEPRESSION IN THEIR CHILDRENS, THE PSYCHOSIS CONSTRUCT PSYCHOSIS TYPES PARTICLE CLOUD CONSTRUCTIONS IN THEIR RECIPIENT INDIVIDUALS, CONSEQUENTLY PRODUCE PSYCHOSE PATIENTS SIMILAR TO THE PARENTS, BIPOLAR PROVIDERS MAY CREATE MANIC DEPRESSIVE DISORDERS AT MOST.
THIS IS THE REASON, THE NORMAL PSYCHES PARENTS CREATE NORMAL THOUGHT CURRENTS FAMILY MEMBERS, ABNORMAL ONES CREATE ABNORMAL CHILDRENS AND STUDENTS, GENIUSES CAN CREATE THE GENIUS, CRIMINALS CREATE CRIMINALS, LEADERS CREATE LEADERS, SPIRITUALS CAN CREATE SPIRITUALS, WORKERS MOSTLY CREATE THE WORKERS, ETC.
THE PHENOMENO OF CHANGING THE RECIPIENT'S PSYCHE, TO BECOME SIMILAR TO PROVIDER ONES, THIS IS DONE, MOSTLY THROUGH CHANGING THE RECIPIENT'S CNS ELECTRON NUCLEON PARTICLE CLOUDS COMPOUNDS CONSTRUCTIONS, SIMILAR TO PROVIDER SUBJECT'S CNS ELECTRON NUCLEON PARTICLE CLOUD COMPOUND CONSTRUCTIONS.
THESE ARE THE PHENOMENONS OF REPLICA – PSYCHE GENESIS, AND PHENOMENONS OF DUPLICATE OR SIMILAR ELECTRON NUCLEON MOLECULAR COMPOUND CONSTRUCTION PHENOMENONS. THE PRODUCED PSYCHE IN RECIPIENT SUBJECTS ARE SIMILAR TO THE PROVIDER SUBJECTS PSYCHE, THE ABNORMAL FP – I. I. – P. Cl PRODUCER SUBJECTS TRANSMIT ABNORMAL PSYCHE AND THOUGHT CURRENTS TO THEIR RECIPIENT CHILDREN, STUDENTS.
IN CONTRAST THE NORMAL FP – I. I. – P. Cl. PRODUCER PARENRTS AND TEACHERS TRANSMIT NORMAL INFORMATIONS AND IMAGES TO THE RECIPIENT CHILDREN AND STUDENTS, AND CREATE NORMAL PSYCHE INDIVIDUALS.

NANO NEUROLOGY OF THE ELECTRONS, NUCLEONS, AND ATOMS

Essential Particle Circulation Systems (E- PCS):

E. PCS = Ex. PCS + I. PCS

THE EXOGENOUS AIR - BORN PARTICLE CLOUD, AND FUNDAMENTAL PARTICLE CURRENTS

THE EXOGENOUS FUNDAMENTAL PARTICLE, AND PARTICLE CLOUD CIRCULATION SYSTEMS (EX. PCS)

Exogenous FP – I.I. – P. Cl. Circulation Systems (EX.- PCS)

THE EXOGENOUS (EXTERNAL ATOM) AIRBORNE FUNDAMENTAL PARTICLE CIRCULATION SYSTEMS (EX – PCS), THE EXOGENOUS AIRBORNE FUNDAMENTAL PARTICLE CLOUD CIRCULATION SYSTEMS, AND THE EXTERNAL ATOM AIRBORNE FUNDAMENTAL PARTICLE INFORMATION AND IMAGE PARTICLE CLOUD CIRCULATION SYSTEMS (EX. – FP – I.I. – P. Cl.), ARE THE PARTICLE CLOUDS CURRENTS IN

SPACE, AND THESE PARTICLE CURRENTS TRAVEL THROUGH AIR BETWEEN DIFFERENT ATOMS, ELECTRONS, NUCLEONS AND NANO UNITS. NANO UNITS FOLLOWING ATTRACTIONS OF THESE PARTICLE CLOUDS, UNDER GRAVITON FORCES, THESE PARTICLES ARE USED IN INTERNAL ELECTRONS NUCLEONS PARTICLE COMPOUND CONSTRUCTIONS OF THE RECIPIENT ELECTRONS, NUCLEONS, AND NANO UNITS.

SOME OF THESE PARTICLE CLOUDS CIRCULATE AND TRAVEL THROUGH BI - DIRECTIONAL CLOSED PATHS PARTICLE CURRENTS BETWEEN SPECIFIC POINTS, ONE PARTICLE CURRENT TRAVEL IN OPPOSITE DIRECTIONS OF THE OTHER PARTICLE CLOUD CURRENTS, AND FUNDAMENTAL PARTICLE CIRCULATION SYSTEMS IN CLOSED CIRCULATION CIRCUITS MOSTLY USD BETWEEN THE GENERAL COMMUNICATIONS BETWEEN THE LIVING THINGS.

SOME OTHER FUNDAMENTAL PARTICLE CURRENTS TRAVEL IN ONE WAY ONE DIRECTIONAL PARRTICLE CLOUD CURRENTS AND PARTICLE CIRCULATION SYSTEMS, MOSTLY CARRY NEEDED PARTICLES FOR NEEDED NANO UNIT SITES. SOME PARTICLE CLOUDS TRAVEL IN BI-DIRECTIONAL CLOSED PARTICLE CLOUD CIRCULATION CIRCUITS SYSTEMS, THESE PARTICLES AND PARTICLE CLOUD CURRENTS CARRY NEEDED PARTICLES EITHER ONE WAY FROM ONE POINT TO OTHER. OR IN TWO WAY PARTICLE CURRENTS, ONE PARTICLE CURRENT IN OPPOSITE DIRECTIONS OF THE OTHER IN CLOSED CIRCULATION SYSTEMS, BETWEEN THE DIFFERENT ATOMS, ALL OVER THE UNIVERSE.

The Indigenous Particle Circulation Systems

(I – PCS)

THESE ARE INTERNAL ATOMS, INTERNAL ELECTRONS, AND INTERNAL NUCLEONS PARTICLE CIRCULATION SYSTEMS, AND INTERNAL NANO UNITS PARTICLE CLOUD CIRCULATION SYSTEMS, THERE ARE NUMEROUS DIFFERENT CLASSES OF FUNDAMENTAL PARTICLE AND PARTICLE – CLOUD CIRCULATION SYSTEMS, INSIDE THE NANO UNIT STRUCTURES:

1 – EXTERNAL ATOMS FUNDAMENTAL PARTICLE CIRCULATION SYSTEMS, PARTICLE CLOUD CIRCULATION SYSTEMS, AND EXTERNAL ATOM F UNDAMENTAL PARTICLE CURRENTS (EX. PCS).
2 – INTERNAL ATOM FUNDAMENTAL PARTICLE CIRCULATION SYSTEMS, AND INTERNAL ATOM FUNDAMENTAL PARTICLE CLOUD CIRCULATION SYSTEMS, AND PARTICLE CLOUD CURRENT (I – PCS).

THE I. – PCS DIVIDE INTO FOLLOWING DIFFERTENT CLASSES:

1 – THE INTERNAL PERIPHERON PARTICLE CIRCULATION SYSTEMS (P – PCS), AND PERIPHERON FUNDAMENTAL PARTICLE CLOUD CIRCULATION SYSTEMS, AND INTER – PERIPHERON FP – I.I. – P. Cl. CIRCULATION SYSTEMS.

2 – THE INTERNAL ELECTRON PARTICLE CIRCULATION SYSTEMS (E – PCS), AND INTERNAL ELECTRON FUNDAMENTAL PARTICLE CLOUD CIRCULATION SYSTEMS, AND INTER – ELECTRON FP – I.I. – P. Cl. CIRCULATION SYSTEMS.

3 – THE INTERNAL NUCLEUS FUNDAMENTAL PARTILE CIRCULATION SYSTEMS (N – PCS), AND INTERNAL NUCLEUS FUNDAMENTAL PARTICLE CLOUD CIRCULATION SYSTEMS, INTER – NUCLEUS FP – I.I. – P. Cl. CIRCULATION SYSTEMS.

4 – THE INTERNAL PROTON FUNDAMENTAL PARTICLE CIRCULATION SYSTEMS, AND PARTICLE CLOUD CIRCULATION SYSTEMS, AND INTER – PROTON FP – I.I. – P. Cl. CIRCULATION SYSTEMS.

5 – THE INTERNAL NEUTRON FUNDAMENTAL PARTICLE CIRCULATION SYSTEMS, AND FUNDAMENTAL PARTICLE CLOUD CIRCULATION SYSTEMS, AND INTER – NEUTRON FP – I.I. – P. Cl. CIRCULATION SYSTEMS.

The Essential Particle Circulation Systems

(E- PCS)

THE E – PCS COMPOSED FROM COMBINATIONS OF THE EX. – PCS, WITH THE I – PCS. WHEN BOTH OF THESE TWO FUNDAMENTAL PARTICLE CIRCULATION SYSTEMS OF EXOGENOUS FUNDAMENTAL PARTICLE CLOUD CURRENTS SYSTEMS CONNECT AND JOIN TO I – PCS, AND CONTINUE AS SINGLE CONNECTED CONTINUOUS FUNDAMENTAL PARTICLE CLOUD CIRCULATION SYSTEMS JOINTLY AS A CLOSED ONE CONNECTED TO EACH OTHER SINGLE CLOSED BI – DIRECTIONAL PARTICLE CLOUD CIRCULATION SYSTEMS.
THIS SINGLE BI- DIRECTIONAL PARTICLE CURRENTS WHICH THE DIRECTIONS OF THESE JOINT TRAVELING PARTICLE CURREENTS, ARE ONE PARTICLE CURRENTS TRAVELING IN OPPOSITE DIRECTIONS OF THE OTHER JOINTLY TRAVELING PARTICLE – CURRENTS, THESE JOINTLY CONNECTED CLOSED CIRCUIT FUNDAMENTAL PARTICLE CLOUD CIRCULATION SYSTEMS, ALL JOINTLY CONNECTED TO EACH OTHER IS CALLED ESSENTIAL FP – I.I. – P. Cl. CIRCULATION SYSTEMS.

General patterns of PCS- Genesis

THE ALL NANO – UNITS, MICRO – UNITS, AND MACRO – UNITS ARE CAPABLE TO CREATE, PRODUCE, AND EMIT FUNDAMENTAL PARTICLE CLOUDS, AND FUNDAMENTAL PARTICLES INFORMATION, IMAGE PARTICLE CLOUDS, AS WELL AS THE FUNDAMENTAL PARTICLE CIRCULATION SYSTEMS (PCS), THE CREATED PCS DIVIDE INTO TWO MAJOR PARTICLE CIRCULATION SYSTEMS, THE EXOGENOUS PARTICLE CLOUD CIRCULATION SYSTEMS (THE EX. – PCS), AND THE INDIGENOUS PARTICLE CLOUD CIRCULATION SYSTEMS (I – PCS).
THE COMBINATONS AND THE JOINT FUNCTION CONNECTIONS OF THE EX. PCS, WITH I- PCS PRODUCE THE ESSENTIAL FUNDAMENTAL PARTICLE CIRCULATION SYSTEMS (E – PCS).
THE PCS IN ALL OF THE ABOVE DIFFERENT SIZE UNITS (NANO UNIT - PCS, MICRO UNIT - PCS, MACRO UNIT - PCS), COMPOSED MOSTLY FROM TWO MAJOR MAIN PARTICLE CIRCULATION SYSTEM: 1 - EX. – PCS, 2 - THE I – PCS.

THE JOINT FUNCTIONAL CONNECTIONS AND PARTICLE CIRCULATION SYSTEMS COMBINATIONS OF THE EX. PCS, WITH THE I – PCS, IN ALL OF THE DIFFERENT NANO – UNITS, MICRO – UNITS, MACRO – UNITS, CONSTRUCT MAIN JOINLTLY FUNCCTIONALLY CCONNECTED ESSENTIAL FUNDAMENTAL PARTICLE CLOUDS CIRCULATION SYSTEMS.

FOLLOWING ARE EXAMPLES FROM LIVING THINGS (MACRO – UNITS) FP – I.I. – P. Cl. EXOGENOUS, INDIGENOUS AND ESSENTIAL PARTICLE CLOUD CIRCULATION SYSTEMS.

PSYCHE GENESIS AND THOUGHT CURRENT GENESIS IN LIVING THINGS CNS ELECTRONS, NUCLEONS

PCS in Macro – Unit Subjects (Living Things)

THE LIVING THINGS EX - PCS, THE LIVING THINGS I – PCS, AND THE E – PCS IN LIVING THINGS

FP- I.I. – P. Cl. STROAGE, AND PSYCHE GENESIS

THE PROVIDER SUBJECT LIVING THINGS EMIT SONIC, LIGHT, THERMAL, ELECTRIC FUNDAMENTAL PARTICLE INFORMATION IMAGE PARTICLE CLOUDS (S. Y. T. E. – FP – I.I. – P. Cl.) FROM THEIR CNS ELECTRONS AND NUCLEONS INTO THE AIR.
THE EMITTED PARTICLE CLOUDS FROM ENVIRONMENTAL SUBJECTS TRAVEL AIRBORNE IN SPACE, UNTIL TO BE CAPTURED BY A RECIPIENT SUBJECTS SENSARY ORGANS ATOMS, UNDER ATTRACTIVE GRAVITON FORCES OF THE SENSARY ORGANS ATOMS ATTRACTIV GRAVITON FORCES.
THE PROVIDER SUBJECT'S FP. - I.I. - P. Cl. UNDER ATTRACTIVE GRAVITON FORCES OF THE SENSARY ATOMS OF THE RECIPIENT SUBJECT FINALLY FROM EX. PCS TRAVEL THROUGH PERIPHERAL ATOM SPACE (P.A.S.) INTO THE RECIPIENT'S SENSARY ATOMS INTERNAL ATOM PCS (I – PCS).

THE EXOGENOUS F.P. – I.I. – P. Cl. CURRENTS, AT INSIDE THE RECIPIENT SUBJECT'S INTERNAL PERIPHERON'S PARTICLE CIRCULATION SYSTEMS (P – PCS), AND INTERNAL NUCLEUS PARTICLE CIRCULATION SYSTEMS (N – PCS), FINDALLY INTER INTO THE RECIPIENT'S CNS INTERNAL ELECTRON'S PARTICLE CIRCULATION SYSTEMS (E – PCS), AND INTERNAL NUCLEON'S PARTICLE CLOUD CIRCULATION SYSTEN (N – PCS), THERE AFTER THE FUNDAMENTAL PARTICLES INFORMATION IMAGE PARTICLE CLOUDS INSIDE SUBSYSTEM UNITS CHEMICA LAB.S COMBINE WITH THE INTERNAL ELECTRONS NUCLEONS PARTICLE COMPOUNDS, AND PRODUCE FP – I.I. – P. Cl. – COMPOUNDS, AND CONSTRUCT THE INTERNAL CNS PARTICLE CLOUD COMPOUND CONSTRUCTIONS OF THE RECIPIENT ATOMS, THROUGH THE USE OF THE PROVIDER SUBJECT'S EX. – FP – I.I. – P. Cl. –COMPOUND CONSTRUCTIONS.

THIS IS STORING PHENOMENON OF THE OUTSIDE INFORMATIONS AND IMAGES IN PARTICLE CLOUD COMPOUND FORMS INSIDE THE CNS ELECTRONS, AND NUCLEONS. WHEN THE STORED S. Y. T. E. – FP – I.I. – P. Cl. INTERACT WITH EACH OTHER THROUGH REGENERATIVE OR DEGENERATIVE A. S. I. FP. Mol. C.I.C., BETWEEN THE DIFFERENT CNS CENTERS DIFFERENT ELECTRONS AND NUCLEONS PARTICLE CLOUD COMPOUNDS, ALSO WHEN THESE INTERACT WITH EXOGENOUS INCOMING S. Y. T. E. – FP – I.I. – P. Cl. O THEY INTERACT WITH EACH OTHER BETWEEN THE DIFFERENT CNS CENTER ELECTRONS AND NUCLEONS, THESE PARTICLE CLOUDS INFORMATIONS AND IMAGES CHEMICAL INTERACTIONS WITH EACH OTHER, BETWEEN ENORMOUS DIFFERENT CNS ATOMS WITH EA OTHER, BIOLOGICALLY FEEL AS THOUGHT CURRENTS AND PSYCHE SYSTEMS. WHICH KNOWN THE SOUL.

AND THIS PHENOMENONS IS PSYCHE – GENESIS AND THOUGHT CURRENT GENESIS PHENOMENONS, THROUGH THE STORING THE OU SIDE INFORMATION AND IMAGES, IN PARTICLE CLOUD FORMS, INSIDE THE CNS ELECTRONS AND NUCLEONS.
THESE ARE ALSO THE PHENOMENONS OF NEO ELECTRON GENESIS, NEO NUCLEON GENESIS, AND NEO PARTICLE CLOUD COMPOUND GENESIS, AND STORAGE OF THE OUTSIDE WORLD'S INFORMATION AND IMAGE PARTICLE CLOUDS, IN PARTICLE CLOUD COMPOUND FORMS, INSIDE THE CNS ATOMS.

THE PCS
Cloud Retrievals and Cloud Responses

(R- Cl.)

COMMUNICATIONS OF DIFFERENT LIVING THINGS, NON LIVING THINGS ELECTRONS, NUCLEONS WITH EACH OTHER

RESPONSE CLOUDS AND RETRIEVAL CLOUDS

NANO – NEUROLOGY IN LIVING THINGS

Cloud Orders (Cl- O) by CNS
Commandeering Electrons and Nucleons in CNS
(C- EN)

IN LIVING THINGS, THE RESPONSE TO OUTSIDE ENVIRONMENTAL INCOMING EXOGENOUS SONIC, LIGHT, THERMAL, ELECTRIC FUNDAMENTAL PARTICLES INFORMATION IMAGE PARTICLE CLOUDS (EX. – s. y. t. e. -FP – I.I. – P. Cl.), IN THE RESPONSE THE LIVING THING CNS COMMANDEERING ELECTRONS, NUCLEONS CENTERS (C- EN), UNDER THE A. S. I. FP. Mol. C.I.C. PRODUCE AUTONOMOUS SEQUENTIAL RESPONSE PARTICLE CLOUDS (R- P. Cl.), AND EMIT FROM SELF THE RESPONE FUNDAMENTAL PARTICLE CLOUDS (R. Cl.) TO THE OUTSIDE SPACE.
FROM THE AIR, THE DIFFERENT SENSARY ORGAN ATOMS OF THE RECIPIENT LIVING THINGS CAPTURE THESE SPACE FLOATING RESPONSE PARTICLE CLOUDS (R. Cl.), AND TRANSIT THE CAPTURED PARTICLE CLOUDS, THROUGH THE RECIPIENT'S INTERNAL BODY PARTICLE CIRCULATION SYSTEMS (PCS) ROUTES THE PARTICLE CLOUDS CARRIED INTO DIFFERENT CNS SENSARY BRAIN CENTERS ELECTRONS AND NUCLEONS.

THERE AFTER THE EX. S. Y. T. E. – FP – I.I. –P. Cl. FROM CNS – SENSARY CENTERS THROUGH INTERNAL CNS – PCS CIRCULATE THROUGH ENTIRE DIFFERENT CNS CENTERS ELECTRONS, NUCLEONS OF THE ALL RELATED DIFFERENT BRAIN CENTERS ELECTRONS, NUCLEONS, AND ATOMS, IN THE MATTER OF THE FRACTIONS OF SECONDS.

THE DIFFERENT CNS ELECTRONS AND NUCLEONS CENTERS, SUCH AS COMMAND CNS ELECTRONS, AND NUCLEONS CENTERS, MOTOR CNS CENTERS ATOMS, THE AUTONOMOUS CNS CENTERS ELECTRONS AND NUCLEONS POPULATIONS, AND RELATED DIFFERENT OTHER CNS CENTERS NANO UNITS, THROUGH THE A. S. I. FP. Mol. C.I.C. IN THE MATTERS OF THE FRACTIONS OF THE SECONDS, THROUGH ONGOING A. S. I. F.P. Mol. C.I.C. BETWEEN THE HUNDREDS OF TRILLIONS DIFFERENT ELECTRONS, NUCLEONS OF TOTAL CNS NANO UNIT POPULATIONS. THROUGH EXPONENTIAL CHEMICAL INTERACTION SPEEDS, WITH WELL COORDINATIONS AND COOPERATIONS OF THE ALL TOTAL BODY ATOMS POPULATIONS, WHICH ONE HELPS TO OTHER, TO ACHIEVE THE ONGOING PARTICLE – TASKS.

THE CHEMICAL INTERACTIONS BETWEEN THE INCOMING PARTICLE CLOUDS MOLECULES, AND PRE-EXISTING INTERNAL ELECTRONS NUCLEONS PARTICLE CLOUDS COMPOUNDS MOLECULAR STRUCTURES OF CNS DIFFERENT CENTERS, UNDER A. S. I. F.P. Mol. C.I.C. SEQUENCES, ACHIEVES NEO-PARTICLE COMPOUND GENESIS WITH INFORMATION IMAGE PARTICLE CLOUDS, AND STORING PARTICLE CLOUDS IN PARTICLE COMPOUND FORMS INSIDE CNS ATOM, PHENOMENON OF NEO –ELECTRON, AND NEO – NUCLEON GENESIS, PHENOMENON OF THE NEO- PSYCHE GENESIS.
AT THE SAME TIMES THESE AUTONOMOUS SEQUENTIAL CHEMICAL INTERACTIONS PRODUCE AND ISSUE FUNDAMENTAL PARTICLE CLOUD ORDERS (O- Cl.) AND PARTICLE CLOUD RESPONSES AGAINST EXTERNAL EVENTS FROM DIFFERENT CENTER CNS ELECTRONS AND NUCLEONS, WHICH THE PRODUCED ORDER – CLOUDS EMITS FROM LIVING THINGS TO THE OUTSIDE ENVIRONMENT SPACE. THESE PARTICLE CLOUDS AS FREE FLOATING SPACE FUNDAMENTAL PARTICLE CLOUDS (FF – FP –Cl.) TRAVEL IN AIR IN EVERY DIRECTION. THESE ARE EMITTED RESPONSE FUNDAMENTAL PARTICLE CLOUDS (R. Cl.).

THE R. Cl. IN SPACE, CIRCULATE THROUGH EXOGENOUS PARTICLE CLOUD CIRCULATION SYSTEMS (EX. PCS) BETWEEN THE DIFFERENT EXISTING PARTICLE CLOUD PROVIDER AND RECIPIENT DIFFERENT LIVING THINGS AND NON LIVING THINGS ELECTRONS, NUCLEONS, AND ATOMS, MOSTLY AIRBORNE, IN SOME CASES THROUGH PERIPHERAL ATOM SPACES PARTICLE CURRENTS PCS, AND THE PARTICLE CLOUD CIRCULATIONS CONTINUE FROM ONE INDIVIDUAL TO THE OTHER, AND IN RECIPIENT INDIVIDUALS INTER INTO DIFFERENT CONTACT RECIPIENT INDIVIDUALS INTERNAL PARTICLE CIRCULATIONS SYSTEMS (I-PCS), TRAVEL THROUGH DIFFERENT BODY

ELECTRONS AND NUCLEONS, REPEAT THE AUTONOMOUS SEQUENCES BACK AND FORTH AS ABOVE, FOLLOWING ANOTHER SEQUENCES, AND ANOTHER PRODUCED R. Cl. THROUGH ENDLESS E – PCS ROUTES.

ABOVE ARE THE GENERAL EXTERNAL BODY, AND INTERNAL HUMAN BODY DIFFERENT PARTICLE CLOUD - PCS ROUTES, THESE PCS ROUTES TRANSMIT THE DIFFERENT NEEDED FUNDAMENTAL PARTICLE CLOUD, AND INFORMATION AND IMAGE PARTICLE CLOUDS, WHICH ARE USED IN ELECTRON, NUCLEON PARTICLE COMPOUND CONSTRUCTIONS OF THE HUMAN LIVING THINGS, AND NON HUMAN EXISTING THINGS ATOMS CONSTRUCTIONS OF THE EARTH.
DURING DAILY COMMUNICATIONS OF THE DIFFERENT ENVIRONMENTAL EXISTING THINGS ELECTRONS NUCLEONS COMMUNICATIONS WITH EACH OTHER, THESE ARE EXOGENOUS PARTICLE CLOUD CIRCULATION SYSTEMS (EX. FP. Cl. – PCS), WHICH CONNECT DIFFERENT INDIGENOUS PARTICLE CLOUD CIRCULATIONS SYSTEMS (I – PCS) OF THE LIVING THINGS ELECTRON, NUCLEONS, DURING THE COMMUNICATIONS OF THE LIVING THINGS AND NON LIVING THINGS ELECTRONS NUCLEONS WITH EACH OTHER, THROUGH USE OF EX. PCS, AND I – PCS, ALL OVER UNIVERSE.

THE STORED S. Y. T. E. – FP – I.I. – P. Cl. – COMPOUNDS INSIDE CNS ELECTRONS, AND NUCLEONS, UNDER THE DEGENERATIVE A. S. I. FP. Mol. C.I.C. AT REVERCE INTERACTION DIRECTIONS, CAN RELEASE THE INFORMATION AND IMAGE PARTICLE CLOUDS TO OUT SIDE OF THE PARTICLE - COMPOUND FORMS,
THE RELEASED INFORMATION AND IMAGE PARTICLE CLOUDS TO OUTSIDE ELECTRONS AND NUCLEONS FROM DIFFERENT CNS CENTERS, IN FREE PARTICLE CLOUD FORMS, BIOLOGICALLY SENSED AS THOUGHT CURRENTS OF THE PAST INCIDENTS, THIS IS ALSO THOUGHT CURRENT GENESIS PHENOMENON.
ABOVE ARE THE CAUSE FOR LIVING THINGS PSYCHE GENESIS AND THOUGHT CURRENT GENESIS PHENOMENONS.

THE BIOLOGICAL NORMAL SENSARY SYMPTOMS, SIGNS AND FEELINGS OF THESE REGENERATIVE AND DEGENERATIVE A. S. I. FP. Mol. C.I.C. PRODUCTIONS OF THE DIFFERENT RELEASED S. Y. T. E. – FP – I.I. – P. Cl., THESE PARTICLE CLOUDS ARE THE INFORMATIONS AND IMAGES OF THE PAST RECORDED INCIDENTS ABOUT ENVIRONMENTAL EVENTS, WHICH THESE DIFFERENT PARTICLE CLOUDS WERE STORED INSIDE THE CNS ELECTRONS NUCLEONS. THEIR RELEASES AND THEIR AUTONOMOUS SEQUENTIAL CHEMICAL INTERACTIONS WITH EACH OTHER BIOLOGICALLY SENSED AND FEELS AS THOUGHT CURRENTS, AND THESE ARE NORMAL PHYSIOLOGICAL SIGNS AND SYMPTOMS OF PARTICLE CLOUD INTERACTIONS WITH EACH OTHER.

IN LIVING THINGS, THE RELEASE OF INFORMATION AND IMAGE PARTICLE CLOUDS SENSED AND CALLED MEMORIZATION PHENOMENON, IN CONTRAST, IN NON LIVE WORLD OF COMPUTER TECHNOLOGIES, UNDER DEGENERATIVE A. S. I. FP. Mol. C.I.C. THE RELEASED PARTICLE CLOUDS CAN BE DEMONSTRATED ON TV SCREENS AND CALLED AS PARTICLE CLOUD RETRIEVAL PROCEDURES.

THE HIERARCHY LAWS AND ORDERS BETWEEN THE MASS UNITS

HIERARCHY SYSTEMS BETWEEN THE NANO UNIT MASSES

ALL OVER IN UNIVERSE, ALWAYS BETWEEN THE SAME SIZE MASS UNITS, SOME SUBCLASSES (FROM SAME MASS UNIT) UNDER HIERARCHY ORDER SYSTEMS CONTROL, ORDER, AND DICTATES AGAINST THE OTHERS FROM SAME KIND MASS UNITS. THIS LAW IS TRUE IN NANO – UNIT MASS CLASSES, SUCH AS THE ELECTRON SIZE AND NUCLEON SIZES MASS UNITS, WHICH BOTH BRIEFLY EXPLAINED AT FOLLOWING PAGES. THIS IS HIERARCHY ORDER AND CONTROL SYSTEMS BETWEEN THE NANO UNIT MASSES.

HIERARCHY SYSTEMS BETWEEN THE MICRO UNIT MASSES

THE HIERACHY ORDER SYSTEMS ALSO IS TRUE IN MICRO UNIT MASSES OF CELL SIZE MASS SPECIES UNITS AS WELL, SOME CELL'S SIZE MICRO UNITS MASSES (MICRO UNIT MASSES) DICTATE, ORDER, CONTROL AGAINST THE OTHER CELL MICRO UNIT MASSES. FOR EXAMPLE, THE CNS MICRO UNIT MASSES AND CELLS ALWAYS AND ALMOST CONTROL THE TOTAL BODY MICRO MASS UNITS CELL POPULATIONS OF THE ALL TOTAL BODY CELL POPULATION. IN ALL LIVING THING SPECIES FROM INSECTS TO THE ANIMAL SPECIES OF THE HUMAN KIND IN PLANET EARTH.
THIS IS THE RULES OF THE HIERARCHY ORDER, AND CONTROL SYSTEMS BETWEEN THE MICRO UNIT MASSES OF THE CELL SIZE GROUPS.

HIERARCHY SYSTEMS BETWEEN THE MACRO UNIT MASSES

REGARDING THE ALL DIFFERENT SPECIES LIVING THINGS MACRO UNIT MASSES IN PLANET EARTH. IN ALL GIVEN SPECIES MACRO MASS UNITS FROM THE INSECTS SIZE MASS UNITS, UP TO THE HUMAN SIZE MASS UNITS, OR TO LARGE SIZE OTHER MASS UNIT LIVING THINGS SUCH AS WHALES OR ELEPHANT SIZE LIVING MACRO MASS UNITS, ALWAYS BETWEEN THE EACH GIVEN SPECIES LIVING THING MASS UNIT MACRO MASS UNIT SPECIES, ALWAYS SOME ANIMALS SPECIES FROM SAME SPECIES AS CHIEF OR KING OR PRESIDENTS OF THE OTHER ANIMALS FROM SAME CLASS ARE DICTATED, ORDERED, AND CONTROLLED ONLY BY ONE BIG NECK ANIMAL. THAT MEANS ONE ANIMAL OR INSECT OR LIVING THINGS ALWAYS UNDER HIERARCHY ORDER SYSTEMS RULES OVER THE OTHER.
THIS IS THE RULES OF THE HIERARCHY ORDER, CONTROL, AND DICTATINGG SYSTEMS ONE MACRO UNIT AGAINST THE OTHER POPULATIONS OF THE WORLD MACRO UNITS. FOR EAXAMPLE EUROPEAN KING MACRO MASS UNIT HUMAN SPECIES CONTROL THE AFRICAN ORIGIN HUMAN MACRO UNIT LIVING THINGS.

RULES OF THE CONTROLS OF THE LARGE SIZE MASS-UNITS AGAINST THE SMALL SIZE MASS UNITS

THE MORE COMMON FORM EXAMPLES FOR ABOVE SYSTEM, IS THE CONTROL, ORDER, AND DICTATION OF LARGER WIEGHT (OR MASS UNIT) OF THE CENTRAL LARGER MASS UNITS UNIVERSE – MASS UNITS, AGAINST THE LESS WEIGHT LESS MASS UNIT, LESS GRAVITON FORCE, LESS ENERGY CONTENTS PERIPHERAL UNIVERSE MASS UNITS. WHICH UNDER THESE UNIVERSAL LARGER AND LESSER MASS UNIT SYSTEMS, ALWAYS THE CENTRAL LARGER MASS UNIT UNIVERSE, BECAUSE OF IT'S LARGER WEIGHT FORCE, LARGER GRAVITON FORCE, AND LARGER ENERGY AMOUNTS CONTENTS ALWAYS CONTROLS, ORDERS AND DICTATES AGAINST THE PERIPHERAL SMALLER MASS UNIT PERIPHERAL UNIVERSE MASS UNITS. ABOVE RULE APPLIES TO ALL DIFFERENT EXISTING UNIVERSES.

THIS RULES AND LAWS OF THE MASSES, ALSO ARE TRUE IN REGARD TO OTHER MASS UNITS, SUCH AS GALAXY SIZE MASS UNITS, AND PLANETARY SIZES MASS UNITS, ETC. AS WELL.

THE RULES AND LAWS OF THE HIERARCHY MASS UNITS POSSESSES DIRECT RELATIONSHIP WITH THE SIZES OF THE MASS UNITS, AS A GENERAL RULE THE LARGER SIZES OF MASS UNITS UNDER THEIR HIGHER ENERGY LEVEL CONTENTS, AND HIGH GRAVITON FORCES AND LARGER MASS UNITS WEIGHTS, ALWAYS THEY CONTROL THE SMALLR SIZES MASS UNITS ALL OVER THE UNIVERSE WITH NO EXCEPTIONS AS THE GENERAL RULES OF THE MASS UNITS.

HIERARCHY SYSTEMS OF ATOMS

CNS Electrons and Nucleon's Hierarchy Systems orders, over Body Organ Electrons Nucleon's Functions, through particle circulation systems

Functional operations of the total body Electron – Nucleon populations under the hierarchy systems of CNS Atoms

NANO – NEUROLOGY OF LIVING THINGS

IN LIVING THINGS OF THE PLANET EARTH, THE PERIPHERAL ORGANS ELECTRONS AND NUCLEONS ALWAYS ACHIEVE ALL OF THEIR PHYSICAL, CHEMICAL, BIOLOGICAL FUNCTIONS, UNDER DIRECT ORDERS, CONTROLS, AND DICTATIONS OF THE DOMINENT ELECTRONS AND NUCLEONS OF THE CNS DIFFERENT CENTERS ELECTRONS NUCLEONS, UNDER THEIR PARTICLE CLOUD ORDERS SYSTEMS.

WHICH THE CNS EMITTED PARTICLE CLOUDS FROM CNS CENTERS ELECTRONS AND NUCLEONS POPULATION AFTER EMISSION TRANSIT THROUGH I – PCS FROM CNS ELECTRONS NUCLEONS TO THE PERIPHERAL BODY ORGANS ELECTRONS, NUCLEONS TOTAL POPULATIONS. THESE PARTICLE CLOUD ORDERS CONTROL THE PERIPHERAL ORGAN'S ELECTRONS AND NUCLEONS FUNCTIONS ACCORDING THE RECEIVED PARTICLE CLOUD ORDERS.

UNDER NANO UNIT MASS CLASSES HIERARCHY LAWS AND ORDERS, THE CNS NANO UNIT ELECTRONS, NUCLEONS, ATOMS, AS DOMINANT COMMANDEERS ELECTRON, NUCLEON, ATOM'S NANO UNIT MASSES THROUGH EMITTED PARTICLE CLOUDS, DECIDE, DICATES, ORDER AGAINST THE PERIPHERAL ORGANS ELECTRON, NUCLEONS POPULATIONS OF THE ENTIREE PERIPHERAL BODY ORGANS ELECTRONS, NUCLEONS, ATOMS AND CELL'S POPULATIONS, WHICH THESE ATOMS AND CELLS OF THE PERIPHERAL BODY ORGANS ALL

MUST DO THEIR PHYSICAL, CHEMICAL, BIOLOGICAL FUNCTIONS ACCORDING THE CNS ATOMS AND CELLS EMITTED PARTICLE CLOUDS ORDERS SYSTEMS.

THE PARTICLE CLOUD CURRENTS ROUTES, THROUGH I – PCS TRAVEL BETWEEN DIFFERENT BODY ORGANS AND SYSTEMS, TRANSIT BACK AND FORTH UNDER BI – DIRECTIONAL AFFERENT AND EFFERENT PARTICLE CURRENTS CIRCULATION SYSTEMS, WHICH ONE PARTICLE CURRENTS RUN IN OPPOSITE DIRECTIONS OF THE OTHER.

THE PARTICLE CLOUD CIRCULATION SYSTEMS CURRENTS (PCS) MOSTLY COMPOSED AND CONTAINS DIFFERENT KINDS OF FLOATING PARTICLES, SUCH AS: ELECTRIC BIO FRIENDLY FUNDAMENTAL PARTICLE CURRENTS, LIGHT FUNDAMENTAL PARTICLE INFORMATION AND IMAGE PARTICLE CLOUDS CURRENTS, SONIC PARTICLES AND PARTICLE CLOUDS INFORMATION IMAGE PARTICLE CLOUD CURRENTS, THERMAL PARTICLES AND OTHER BIO FRIENDLY PARTICLE CLOUD INFORMATION AND IMAGE PARTICLE CLOUD CURRENTS AND PARTICLE CIRCULATION SYSTEMS (PCS).
THERE ARE MANY OTHER TYPES KNOWN AND UNKNOWN PARTICLES AND PARTICLE CLOUDS ALL FLOW THROUGH THESE PCS ROUTES FROM ONE ORGAN ELECTRONS, NUCLEONS, TO OTHER BODY ORGANS, AND BODY SYSTEMS DIFFERENT ELECTONS AND NUCLEONS PARTICLE INTELLIGENCE SYSTEM CENTERS (PIS), ALL PARTICLE CIRCULATIONS ARE BI – DIRECTIONAL BETWEEN THE COMMANDER ELECTRONS NUCLEONS CENTERS, AND RECESSIVE PERIPHERAL ORGANS AND SYSTEMS ELECTRONS, NUCLONS POPULATION NANO UNIT MASSES.

IN LIVING THINGS, THE LOWER LEVEL PERIPHERAL ORGAN'S RECESSIVE ELECTRONS AND NUCLEONS POPULATIONS, ONLY POSSESS FEED BACK PARTICLE CLOUD ISSUANCE CAPABILITIES AT MOST, THAT MEANS THE PERIPHERAL ORGANS ELECTRONS AND NUCLEONS POSSESSES THE INFORMATION AND IMAGE PARTICLE CLOUD EMISSIONS CAPABILITIES, AND PARTICLE CLOUD REPORTING CAPABILITIES, ABOUT THE EXISTING SITUATIONS THAT EXIST IN PERIPHERAL BODY ORGANS FUNCTIONS IN REGARD TO CHEMICAL TASKS, PHYSICAL FUNCTIONS, AND BIOLOGICAL FUNCTIONS STATES OF THE PERIPHERAL ORGANS, NANO UNITS, ATOMS, AND CELLS, ALL GET REPORTED IN RETROGRADE INFORMATION PARTICLE CLOUDS FORMS, THROUGH THE PCS CURRENTS BACK TO THE CNS ELECTRONS AND NUCLEONS DIFFERENT COMMANDER SYSTEM CENTERS, FOR THEIR INFORMATIONS, ANALYSIS, AND ORDER EMISSIONS IN OTHER DECISIONS.

THE PERIPHERAL BODY ORGANS RECESSIVE ELECTRONS AND NUCLEONS ONLY POSSESSES EXCEPTIONAL MINOR LOCAL AUTONOMIES TO CORRECT OR REFUSE TO DO ISSUED ORDERS, WHICH THEY FEEL THOSE ARE OBVIOUSLY INCORRECT, AND PREVIOUSLY FROM BEFORE THEY HAVE BEEN PRE – PROGRAMMED ABOUT POSSIBILITIES OF THE OCCURANCES OF THESE WRONG PARTICLE CLOUD ORDERS.

THE CENTRAL INTELLIGENCE SYSTEMS (CIS), OF THE PERIPHERAL ORGANS ELECTRONS AND NUCLEONS PRODUCE FP – I.I. – P. CI. ABOUT ONGOING EXISTING SITUATIONS ABOUT PERIPHERAL ORGANS FUNCTIONS ABOUT ALL EXISTING TASKS, EITHER OCCURRING NORMALLY, OR HAPPENNING AT INCORRENT PATTERNS.

THESE PARTICLE CLOUD REPORT FOLLOWING PRODUCTION, TRANSIT FROM PERIPHERY TO CNS CENTERS ELECTRONS AND NUCLEONS, THROUGH PCS, WHICH THEY TRAVEL WITH SPEEDS OF LIGHT PARTICLES, AND PROVIDING THE PERIPHERAL ORGANS NORMAL OR ABNORMAL INCIDENTS, INJURIES, DISEASES, PHYSICAL, CHEMICAL, BIOLOGICAL INFORMATIONS IMAGES IN PARTICLE CLOUD FORMS, ALL PERPORTED TO CNS ELECTRONS NUCLEONS, FOR THEIR INFORMATIONS, DECISIONS, AND ORDERS. UNDER HIERARCHY ELECTRON NUCLEON SYSTEMS.

CNS Electrons and Nucleons Control total Body Organ's Electrons, Nucleon's Functions Under Hierarchy Control Systems and Particle Clouds Order Operations systems, through total body PCS

IN LIVING THINGS, ALL EXISTIG BODY ORGAN'S FUNCTIONS TAKE PLACE ACCORDING THE PARTICLE CLOUD ORDERS, WHICH HAVE BEEN EMITTED AND DECIDED, THROUGH THE HIGHER CENTRAL INTELLIGENCE ELECTRONS AND NUCLEONS SYSTEM CENTERS ORDERS AND DECISIONS.
MOSTLY THE CNS ELECTRONS NUCLEONS DECIDE, PRODUCE AND EMIT THE PARTICLE CLOUDS, ADDITIONALLY THE BODY ORGANS AND BODY SYSTEM'S CENTRAL INTELLIGENCE SYSTEMS CENTER'S COMMANDEERING ELECTRONS AND NUCLEONS CENTERS, ALSO DECIDE, EMIT AND ORDER PARTICLE CLOUDS PRODUCTIONS UNDER THE REGENERATIVE A. S. I. FP. Mol. C.I.C.
THE EMITTED PARTICLE CLOUDS FROM DIFFERENT CENTERS OF THE CNS, OR EMITTED PARTICLE CLOUDS FROM BODY ORGANS CENTRAL INTELLIGENCE SYSTEM CENTERS ELECTRONS AND NUCLEONS, THEREAFTER THE EMISSION TRAVEL THROUGH INDIGENOUS PARTICLE CIRCULATION SYSTEMS (I- PCS) TO THE ALL PERIPHERAL TOTAL BODY ELECTRONS AND NUCLEONS POPULATIONS, UNDER THE WELL ORGANIZED SPECIFIC PARTICLE CLOUD CIRCULATION SYSTEMS.
THE PARTICLE CLOUDS AT TARGET PERIPHERAL ORGANS CAUSE THE TOTAL BODY ELECTRONS AND NUCLEONS POPULATIONS, (WHICH THE TOTAL BODY ELECTRON, NUCLEON POPULATIONS NUMBERS IN EARTHS LIVING THINGS BODY, RELATIVELY EXCEEDS TEN TO THE POWER OF FIFTY ELECTRON AND NUCLEON POPULATIONS TOTALLY), PERFORM THEIR BIOLOGICAL, CHEMICAL, PHYSICAL FUNCTIONS ACCORDING THE PARTICLE CLOUD ORDERS.

ALL OF THESE TOTAL BODY ELECTRONS AND NUCLEONS POPULATIONS ALL MUST FUNCTION EXACTLY ACCORDING THE ORDERS AND INSTRUCTIONS OF THE PRODUCED AND ORDERED PARTICLE CLOUDS OF THE HIGHER CENTRAL INTELLIGENCE SYSTEM CENTERS OF THE BODY ORGANS AND BODY SYSTEMS.
ALL OF THESE ORDER PARTICLE CLOUDS (O – P. Cl.) HAVE BEEN DECIDED, PRODUCED, EMITTED AND ORDERED THROUGH SUPERIOR CNS ELECTRON, NUCLEON'S COMMAND CENTER SYSTEMS ELECTRONS, AND NUCLEONS, OR THROUGH THE OTHER BODY SYSTEMS AND BODY ORGAN'S PARTICLE INTELLIGENCE SYSTEM CENTERS (PIS) ATOMS.

THE ORDER PARTICLE CLOUD (O – P. Cl.) TRAVEL TO PERIPHERAL ORGAN'S ELECTRONS NUCLEONS POPULATIONS BY PCS, AND THERE THE PARTICLE CLOUDS INTERACT WITH PERIPHERAL BODY ORGANS AND BODY SYSTEMS INTRNAL ELECTRON, INTERNAL NUCLEONS PARTICLE COMPOUNDS, UNDER THE A. S. I. FP. Mol. C.I.C. AND CONTROL THE ALL AND ENTIRE TOTAL BODY ORGANS AND BODY SYSTEMS PHYSICAL, CHEMICAL, BIOLOGICAL FUNCTIONS, AND TASK OPERATIONS.
THE BODY ELECTRONS NUCLEONS, ATOMS, CELLS, AND BODY ORGANS AND SYSTEMS FUNCTIONS ALL DONE UNDER CNS PRODUCED AND ORDERED PARTICLE CLOUD INSTRUCTIONS, ADDITIONALLY THE BODY ORGANS AND SYSTEMS FUNCTIONS ALL CONTINUOUSLY CHECKED, CONTROLLED, REGULATED, AND WATCHED IN EVERY PASSING SECONDS THROUGH THE COMMANDEERING CNS ELECTRONS AND NUCLEONS, BY PARTICLE CLOUD REPORTS UNDER I – PCS IN ALL TIMES ACCORDINGLY.

THE VOLUNTARY FUNCTIONING CNS ELECTRONS AND NUCLEON CENTER SYSTEMS
THE AUTONOMOUS NON VOLUNTARY FUNCTIONING CNS ELECTRONS AND NUCLEONS CENTER SYSTEMS

THE CNS ELECTRONS AND NUCLEONS CENTERS DIVIDE INTO TWO MAJOR SEPARATE SUBCLASS SYSTEM CENTERS:
THE CNS ELECTRONS NUCLEONS EITHER THEY ARE PART OF THE VOLUNTARY CNS CENTERS, WHICH THEIR PARTICLE CLOUDS EMISSION, RELEASES AND OPERATING SYSTEMS, ALL PARTICLE CLOUD ORDERS FUNCTIONS IN DIFFERENT BODY ORGANS AND SYSTEMS, ALL ARE CONTROLLED DIRECTLY UNDER VOLUNTARY OPERATING CNS ELECTRONS AND NUCLEONS CENTER SYSTEMS FUNCTIONAL OPERATIONS. THE RELEASE OF VOLUNTARY PARTICLE CLOUDS ALL DECIDED BY THE CNS CENTERS ELECTRONS AND NUCLEONS, WHOSE FUNCTIONS CONTROLLED BY VOLUNTARY CNS ELECTRONS AND NUCLEONS OF THE LIVING THING.

IN CONTRARY THE OTHER SUBCLASS OF THE CNS ELECTRONS AND NUCLEON'S FUNCTIONS ARE AUTONOMOUS AND THEIR FUNCTION ARE UNDER NON VOLUNTARY CNS AUTONOMOUS FUNCTIONING ELECTRONS, AND NUCLEONS DECISIONS AND ORDER SYSTEMS, AND THESE CNS CENTER ELECTRONS AND NUCLEONS PRODUCE PARTICLE CLOUDS, AND THESE PARTICLE CLOUDS OPERATE ALL BODY ORGANS AND BODY SYSTEMS WHICH THEIR PHYSICAL, CHEMICAL, BIOLOGICAL FUNCTIONS ARE AUTONOMOUS, AND CONTROLLED IN ALL TIMES, UNDER THE NON – VOLUNTARY CNS ELECTRONS AND NUCLEONS.

ABOVE TWO DIFFERENT CNS ELECTRONS AND NUCLEONS AND THEIR EMITTED PARTICLE CLOUDS FUNCTION ENTIRELY ARE DIFFERENT FROM EACH OTHER. THE VOLUNTARY SUBCLASS CNS ELECTRONS, NUCLEONS EMIT VOLUNTARY PARTICLE CLOUDS, AND ARE UNDER VOLUNTARY DECISION MAKING CENTRAL INTELLIGENCE SYSTEM CENTERS ATOMS, THE EMITTED PARTICLE CLOUDS ORDERS FROM THESE CENTERS, TRANSMIT THROUGH VOLUNTARY PARTICLE CLOUDS, THROUGH PCS TO THE SPECIFIC VOLUNTARY FUNCTIONING BODY ORGANS AND SYSTEMS.

THE VOLUNTARY FUNCTIONING BODY ORGAN AND SYSTEMS ALL FUNCTIONS UNDER PARRTICLE CLOUD ORDERS OF THE VOLUNTARY CNS CENTERS ELECTRONS, NUCLEONS ORDERS. FOR EXAMPLE, THE MUSCULOSKELETAL BODY ORGANS AND SYSTEMS ARE UNDER THE ORDERS OF VOLUNTARY CNS ELECTRONS AND NUCLEONS.

THE OTHER MAJOR SUBCLASS CNS ELECTRONS NUCLEONS ARE AUTONOMOUS SUBCLASS ATOMS CENTERS.
THE AUTONOMOUS SUBCLASS CNS ELECTRON, NUCLEONS CENTERS EMIT DIFFERENT ORDERS AND TYPES FUNDAMENTAL PARTICLE CLOUDS, WHICH THESE PARTICLE CLOUDS PRODUCE AND CAUSE SPECIFIC PATTERNS CONTINUOUS AUTONOMOUS PHYSICAL, CHEMICAL, AND BIOLOGICAL FUNCTIONS, IN RECIPIENTS BODY ORGANS AND SYSTEMS ELECTRONS AND NUCLEONS. A FEW EXAMPLES FROM AUTONOMOUSLY FUNCTIONING BODY ORGANS AND SYSTEMS ARE SUCH AS: GIS, GUS, CARDIO RESPIRATORY AND VASCULAR SYSTEMS, ETC.

BODY ORGAN'S PCS OPERATIONS UNDER CNS PARTICLE CLOUDS

(CNS, V-PCS) Vertical Particle Circulations Systems

THE ORDERED AND EMITTED PARTICLE CLOUDS FROM CNS COMMAND CENTER'S ELECTRONS, AND NUCLEONS (O- Cl.) TRAVEL THROUGH INDIGENOUS PARTICLE CIRCULATION SYSTEMS (I – PCS), DIRECT TO RELATED BODY ORGAN'S AND BODY SYSTEM'S INDEPENDENT PARTICLE INTELLIGENCE CENTER SYSTEMS (PIS), AND DELIVER THE CNS ORDERED PARTICLE CLOUDS (O - P Cl.) DIRECTLY TO THE PERIPHERAL BODYS ORGAN'S PIS, AND PERIPHERAL BODY SYSTEM'S INDEPENDENT PARTICLE INTELLECTUAL SYSTEMS CENTERS (PIS). THESE TYPES I- PCS ALMOST ARE EITHER DESCENDANT VERTICAL PARTICLE CIRCULATION SYSTEMS (V-PCS), OR ASCENDANT VERTICAL PARTICLE CIRCULATION SYSTEMS (V-PCS).

DIFFERENT CNS CENTERS ELECTRONS, NUCLEONS THROUGH EMITTED PARTICLE CLOUDS ORDERS (O- P Cl.) CAUSE BIOLOGICAL, PHYSICAL, CHEMICAL FUNCTIONS OF THE BODY SYSTEMS AND BODY ORGANS ENTIRE TOTAL ELECTRONS NUCLEONS POPULATIONS UNDER THE O – P Cl. ORDERS.

V – PCS OF BODY SYSTEMS AND BODY ORGANS

THE EMITTED AND ORDERED PARTICLE CLOUDS (O – P Cl.) FROM BODY SYSTEM'S PARTICLE INTELLIGENCE SYSTEMS CENTERS (S – PIS) ALSO THE EMITTED ORDER – PARTICLE CLOUDS (O- P Cl.) FROM BODY ORGAN'S PARTICLE INTELLIGENCE SYSTEM CENTERS (O – PIS), THE BOTH EMITTED ORDER PARTICLE CLOUDS FROM O – PIS AND S - PIS TRAVEL EITHER DESCENDANT VERTICAL INDIGENOUS PARTIC CIRCULATIONS SYSTEMS PATTERNS, AND CAUSE OPERATIONS OF THE TOTAL PERIPHERAL BODY ORGAND ELECTRONS AND NUCLEONS WHICH EXIST AND LIVE IN DIFFERENT ORGANS AND SYSTEMS OF THE BODY.

ALSO THE TOTAL PERIPHERAL BODY ORGAN'S ELECTRONS AND NUCLEONS PRODUCE, AND EMIT ABOUT EXISTING INFORMATIONS AN IMAGES WHICH ARE GOING ON IN PERIPHERAL BODY ORGANS BIOLOGICAL, PHYSICAL AND CHEMICAL FUNCTIONS IN PARTICLE CLOU INFORMATION AND IMAGE FORMS AND TRANSMIT ALL OF THOSE REPORT INFORMATION PARTICLE CLOUDS (R P Cl.) THROUGH THE REVERSE DIRECTION RETRO-GRADE VERTIVAL ASCENDANT PCS (VA – PCS) BACK TO THE HIGHER COMMANDER PARTICLE INTELLECTU ELECTRONS AND NUCLEONS AT CNS CENTERS DECISION MAKING CENTERS. UNDER DESCENDING PCS AND ASCENDING PCS THE TOTA BODY ELECTRONS AND NUCLEONS POPULATIONS COMMUNICATE WITH EACH OTHER THROUGH I – PCS.

UNDER THESE ASCENDING OR DESCENDING PCS FROM CNS ELECTRONS, NUCLEONS, TOWARD THE PERIPHERAL BODY ORGAN'S, AND BODY SYSTEM'S TOTAL BODY ELECTRONS AND NUCLEONS POPULATIONS, THE CNS ELECTRONS AND NUCLEONS EMITTED PARTICLE CLOUD ORDERS CONTROL THE ENTIRE TOTAL BODY POPULATIONS ELECTRONS AND NUCLEONS POPULATIONS PHYSICAL, CHEMICAL, BIOLOGICAL FUNCTIONS.

EXAMPLE OF VERTICAL, AND HORIZENTAL PCS FOR ANIMAL'S CARDIOVASCULAR SYSTEMS AND ORGAN

THE AUTONOMOUSLY ISSUED PARTICLE CLOUD ORDERS FROM THE AUTONOMOUS CNS ORDER CENTERS, IN ORDER TO REGULATE AND OPERATE AUTONOMOUSLY THE CARDIAC AND VASCULAR FUNCTIONS, AS WELL AS THE OTHER ALL NUMENOUS AUTONOMOUSLY FUNCTIONING BODY ORGANS, AND SYSTEMS FUNCTIONS SUCH AS GIS, GUS, CNS, PULMONARY RESPITATION SYSTEMS OR DEFLATION-INFLATION BIO –DEVICES, CARDIAC FLUID PUMPER BIO-DEVICES, ETC. THE CNS AUTONOMOUS CENTERS ELECTRONS, NUCLEONS, AND PERIPHERAL AUTONOMOUS CENTERS ELECTRONS AND NUCLEON'S EMIT AND PRODUCE PARTICLE CLOUDS ORDERS (O – P Cl.). THESE P Cl. -O TRAVEL FROM AUTONOMOUS CNS CENTERS DIRECT TO THE CARDIAC ORGANS PARTICLE INTELLIGENCE SYSTEM CENTERS (SUCH AS CARDIAC AURICULAR HES NODAL PARTICLE INTELLIGENCE SYSTEM CENTER), OR ANY OTHER AUTONOMOUS FUNCTIONNING ORGANS AND SYSTEMS PARTICLE INTELLIGENCE SYSTEMS CENTERS, AND OPERATE THE DIFFERENT BODY ORGANS AND BODY SYSTEMS TASKS.

THE CARDIAC AURICULAR HESS - NODE PARTICLE INTELLIGENCE SYSTEM (H - PIS) CENTER IS IN CHARGE OF CARDIAC ORGAN'S ELECTRONS, NUCLEON'S PHYSICAL, CHEMICAL, BIOLOGICAL FUNCTIONS OPERATIONS, UNDER THE DESCENDING CARDIAC PCS, AND ASCENDING HEART'S PCS, THE PARTICLE CLOUD –ORDERS WHICH HAS BEEN ISSUED AND EMITTED FROM COMMAND AUTONOMOUS AND PERIPHERAL AUTONOMOUS INTELLIGENCES SYSTEMS CENTERS, ALL WITH EACH OTHERS UNDER THE WELL COORDICATED AND COOPERATED INTELLIGENCE SYSTEM CENTERS FUNCTION, ACHIEVE GREAT CARDIAC FUNCTIONAL TASKS, THROUGH THE FUNDAMENTAL PARTICLE'S PRECISSIONS ACCURACIES, WITH NO MISTAKES 24/ 7 DAY AND NIGHT CORRECTLY ACCHIEVED WITH NO MISTAKE MANY STROKES PER MINUTES DEPENDING TO THE DIFFERENCES IN SPECIES, WITHOUT MISTAKES.
(THE ABOVE IS UNDER THE ELECTRON'S NUCLEON'S VERTICAL PCS, AND PIS ORDER SYSTEMS),

IN CONTRARY UNDER THE HORIZONTAL PCS, AND UNDER THE DIFFERENT ORGAN'S PIS, THERE ARE ADDITIONAL ANOTHER PCS, AND PIS WHICH AFFECT THE ORGAN'S FUNCTIONS, FOR EXAMPLE IN CARDIAC FUNCTIONS EXAMPLE, THE CARDIA FUNCTION ALSO IS INFLUENCES UDER THE PARTICLE CLOUDS ORDERS WHICH ARE COMING FROM THE ANOTHER BODY ORGANS.
FOR EXAMPLE, PLACING A VASO-DIALATOR SUCH AS NTG UNDER TONGUE FROM GIS (OR OTHERS FROM ANOTHER ORGANS), THE PARTICLE CLOUDS HORIZONTALLY DIRECT TRAVEL FROM GIS OR OTHER SYSTEMS TO HEART, AFFEXCTS FUNCTIONS DIRECTLLY, THESE ANOTHER ORGANS PARTICLE CLOUDS, WHICH ORDERED FROM ANOTHER ORGAN'S PIS, SUCH AS GIS CLOUDS ABOUT NTG, THROUGH HORIZONTAL PCS, TRAVEL HORIZONTALLY DIRECT FROM GIS TO CAEDIO-VASCULAR SYSTEMS PIS, REPORT THE PARTICLE INFORMATION AND IMAGES PARTICLE CLOUDS ORDERS FROM GIS TO CARDIAC SYSTEMS, WITH SPEEDS OF LIGHT, CAUSE CARDIAC VASODIALATION AND ANGINAL CARDIA PAIN RELEAF WITHIN SECONDS (UNDER HORIZONTAL PCS AND PIS).
(THESE ARE HORIZONTAL OPERTIONS OF THE PARTICLE CLOUDS TRANSITS FROM ONE ORGAN PIS TO OTHER ORGAN'S PIS, THROUGH THE HORIZONTAL PIS, AND PCS).

THE PIS OPERATIONS FOR ALL OTHER BODY ORGANS ELECTRONS NUCLEONS FUNCTIONS, AND I- PCS ALL FOLLOW THE CARDIAC SYSTEM'S EXAMPLE THOROUGHLY WITH NO EXCEPTIONS ALL THE SAME, THROUGH THE AUTONOMOUS CNS PARTICLE CLOUD ORDERS, THE OTHER BODY SYSTEMS AND BODY ORGANS O – P Cl. SUCH AS: THE GI S, THE GUS, THE RESPIRATORY SYSTEM AUTONOMOUSLY FUNCTIONING BIO DEVICES, UNDER THE AUTONOMOUS SYSTEMS CENTERS, AND OTHER BODY ORGAN AND BODY SYSTEMS AUTONOMOUS OPERATIONAL SYSTEMS DIRCTLY TRANSIT PARTICLE CLOUD ORDERS TO THE RELATED ORGANS AND SYSTEMS, AND PRODUCE THE ORGAN'S PHYSICAL, CHEMICAL, BIOLOGICAL FUNCTIONS OF THE ALL PERIPHERAL BODY ORGANS AND SYSTEMS TOTAL ELECTRONS NUCLEONS BODY POPULATIONS, DAY AND NIGHT WITH NO INTERRUPTIONS AND NO MISTAKES.

IN INVOLUNTARY BODY ORGANS AND BODY SYSTEMS INTELLECTUAL SYSTEMS MOSTLY FUNCTIONS UNDER AUTONOMOUS CNS CNTERS EMITTED PARTICLE CLOUD ORDERS, AND ALL FOLLOW THE TWO BI – DIRECTIONAL AUTONOMOUS ASCENDING AND DESCENDING PARTICLE CIRCULATION SYSTEMS PATTERNS. IN REGARD TO THE AUTONOMOUSLY FUNCTIONING ORGANS AND SYSTEMS.
ADDITIONALLY, IN CONTRAST THERE ARE SOME OTHER ORGANS AND SYSTEMS WHICH THEIR FUNCTIONS ARE UNDER VOLUNTARY CNS PARTICLE CLOUD ORDER SYSTEMS. SUCH MUSCULO SKELETHAL SYSTEMS FUNCTIONS ARE UNDER DIFFERENT CNS ORDERS, AND THEY ARE UNDER THE VOLUNTARY CNS ELECTRONS, NUCLEONS PARTICLE CLOUD ORDER SYSTEMS OPERATIONS AS WELL. THESE SYSTEMS PCS ALSO ARE SIMILAR TO THE AUTONOMOUS PCS OPERATIONS WITH NO EXCEPTIONS ALL SIMILARLY, THE VOLUNTARY ASCENDING P Cl. ORDER SYSTEMS, THE VOLUNTARY DESCENDING P. Cl. – 0 CIRULATION SYSTEMS ORDERS., ETC.

Horizontal Particle Circulation Systems (H – PCS)

THE HORIZONTAL PARTICLE CIRCULATION SYSTEMS (H – PCS) ARE DIFFERENT TYPE PARTICLE CIRCULATIONS SYSTEMS THAT OCCUR BETWEEN THE PERIPHERAL BODY ORGAN'S PIS, AND BETWEEN THE DIFFERENT BODY SYSTEM'S PIS.
THE H – PCS ALSO CONNECTS DIFFERENT BODY SYSTEM'S, DIFFERENT BODY ORGAN'S PIS, TO EACH OTHER.
IT ALSO CONNECTS THE DIFFERENT ORGAN'S PIS OF A GIVEN BODY SYSTEMS PIS TO EACH OTHER THROUGH THE PCS.

FOLLOWING EXAMPLE DESCRIBE THE DIFFERENT FUNCTIONS OF THESE ORGAN'S PIS HOW THEY FUNCTION WHEN ONE CONNECT TO OTHER:
PLACE NITROGLYCERINE DROP UNDER TONGUE AT UPPER GI S – PIS CENTERS. IMMEDIATELY THE GI S – PIS EMIT AND REPORT, THROUGH HORIZONTAL – PCS, THE EXISTEENCES OF THE NTG PARTICLE INFORMATION AND IMAGE CLOUDS, UNDER THE TONGUE

IMMEDIATELY TRANSIT THROUGH H- PCS FROM GI S PIS CENTERS TO THE INFORMATIONS OF THE CARDIAC AURICULAR NODE PIS – HES SYSTEMS CENTERS,
UNDER THE PERCEIVED NTG – PARTICLE CLOUD INFORMATIONS EXISTENCES UNDER TONGUE –PIS REPORTS FROM GI S - PIS, THE CARDIAC HES PIS ORDERS PROPER CARDIAC MODALITIES, SUCH AS DIALATION OF THE CARDIAC VESSELS, BETTER CARDIAC CIRCULATION SYSTEMS FUNCTIONS, AND RELIEF OF CHEST PAIN, IF IT IS EXISTED.
THE HORIZONTAL PARTICLE CIRCULATION SYSTEMS ALSO ALL ARE BIDIRECTIONAL HORIZONTAL PARTICLE CIRCULATION SYSTEMS, BETWEEN THE UPPER GI S PIS CENTERS, AND CARDIAC – PIS CENTERS, WHICH THE EMITTED PARTICLE CLOUDS WITH SPEEDS OF THE LIGHT, THE PARTICLE CLOUD CIRCULATION BETWEEN TWO RELATED ORGANS TAKE PLACE, AND NECESSARILY NEEDED JOB IS DONE, WITH IN THE MATTERS OF SECONDS THROUGH THE SPEEDS OF THE LIGHT, MINUS TRANSIT ROUTE RESISTENCES.

ANOTHER EXAMPLES FROM H - PCS:

1- SAME BODY SYSTEM H – PCS: DRINKING WATER AND TAKING FOOD IN SOME INDIVIDUALS MAY TRIGGER PARTICLE CLOUD TRANSMISSION FROM UPPER GIS PIS CENTERS TO REPORT THE INFORMATIONS BY PARTICLE CLOUD ROUTES TO THE DISTAL COLON PIS CENTERS, AND IN DISTAL COLON THESE PARTICLE CLOUD ORDERS MAY TRIGGER PHYSIOLOGICAL FUNCTIONS IN GI S. AT SOME INDIVIDUALS, IT IS THE SAME, TAKING THE WATER IN SOME INDIVIDUALS MAY TRIGGER PARTICLE CLOUD INFORMATION PCS INFORMATION TRANSFER FROM UPPER GU S CENTERS TO BE EMITTED AND CARRIED TO DISTAL GU S ELECTRONS AND NUCLEONS AND CAUSE SIGNS AND SYMPTONS OF THE URINARY FUNCTIONS. AND TRIGGER THE PHYSIOLOGICAL FUNCTIONS IN UG S AT SOME INDIVIDUALS,

2 - IN ELDERLY BECAUSE OF MALFUNCTIONS OF THE PCS, AND PIS, MAY CAUSE SYMPTOMS AND SIGNS MIX UPS BETWEEN THE TWO ADJASCENT BODY SYSTEMS AND BODY ORGANS, IN ABOVE EXAMPLE TAKING WATER MAY CAUSE FALSE TRANSIT PARTICLE CLOUDES TO DISTAL COLON (IN STEAD OF DISTAL GU S) THROUGH MALFUNCTIONING AND SENSE OF DISTAL COLON URGENT FUNCTIONAL NEEDS, OR TAKING FOOD BY MOUTH THROUGH FLASE PCS CIRCUITS TRANSIT TO DISTAL GU S, AND CAUSE SENSE OF DISTAL U S FUNCTIONS (IN STEAD OF GI S FUNCTIONS).

MALFUNCTIONS OF THE PIS AND
MALFUNCTION THROUGH FALSE PCS AND FALSE PARTICLE CLOUD TRANSIT SYSTEMS

PARTICLE CLOUD DISORDERS

CLASSIFICATION OF INFORMATION-IMAGE PARTICLE CLOUDS:

FP – I.I. – P. Cl. divide into two main classes:

1 - Normal Particle Clouds, cause Normal Psyche, and normal body Function.

2 - Abnormal Particle clouds cause abnormality of psyche and abnormal body systems.

THE FALSE PARTICLE INFORMATION CLOUD TRANSFERS TO WRONG ORGANS, ALSO MALFUNCTIONING OF THE PARTICLE INTELLIGENCE SYSTEMS OF THE DIFFERENT BODY ORGANS AND BODY SYSTEMS ALL CAN PRODUCE DISORDERS. ALSO FALSE PCS CURRENTS TRANSFERS INTO A GIVEN BODY ORGAN, OR ABNORMAL PARTICLE CLOUD INFORMATION TRANSITS INTO WRONG BODY ORGAN LOCATIONS AND WRONG BODY SYSTEMS LOCATIONS ALL CAN PRODUCE FUNCTION DISORDERS IN ORGANS AND SYSTEMS DISORDERS, PRODUCTIONS OF ABNORMAL SIGNS AND SYMPTOMS IN BODY ORGANS AND SYSTEMS, GENESIS OF THE DISEASES.
THESE ARE SUBJECTS OF THE AUTHORS ANOTHER TEXTS.
THE WRONG INFORMATION AND IMAGE PARTICLE CLOUDS PCS, ALSO THE PIS MALFUNCTIONS AND DISORDERS UNDER DIFFERENT DISEASE SITUATIONS CAUSE AND PRODUCE ENORMOUS DIFFERENT KINDS SIGNS, SYMPTOMS, FALSE SENSARIES, PAINS, OR MOTOR ABNORMALITIES, AND FEELINGS IN DIFFERENT BODY ORGANS, IN SOME CASES EVEN THE ORGANS MAY ARE NORMAL. FALSE OR ABNORMAL PARTICLE CLOUD TRANSFERS ALSO CAN CAUSE ORGAN ABNORMALITY DEVELOPEMENTS IN NON RELATED BODY REGIONS. EVEN MANY PRESENTLY KNOWN DISORDERS SUCH AS; IRRITABLITIES IN BOWL, GIS, STOMACH, GUS, CHEST SENSATIONS, CHEST PAINS, FALSE BREATHING DISORDERS AND SHORTNESS OF BREATHING SENSATIONS, EVEN FALSE PARALYSIS, AND UNABILITIES FOR MOVEMENTS, ETC. MANY OF THESE ARE PRODUCED UNDER THE PARTICLE CLOUDS DISORDERS CAUSES.

PARTICLE INTELLIGENCE SYSTEM CENTERS OF ATOMS, ELECTRONS, NUCLEONS

THE SUBJECT'S CLOUD (S. C.)

The Particle Cloud is Subject's- Copy

Phenomenon of Particle Cloud genesis

THE PHENOMENON OF THE FUNDAMENTAL PARTICLE INFORMATION IMAGE PARTICLE CLOUD GENESIS

THE COMMON KNOWN BIO FRIENDLY FUNDAMENTAL PARTICLES, PARTICLE CLOUDS, FUNDAMENTAL PARTICLES INFORMATIONS AND IMAGES PARTICLE CLOUDS (FP – I.I. – P. Cl.) IN PLANETS EARTH'S ATMOSPHERE AND SURFACE ARE THE DIFFERENT CLASSES FROM LIGHT FUNDAMENTAL PARTICLES, SONIC FUNDAMENTAL PARTICLES, AND BIO FRIENDLY ELECTRIC AND THERMAL FUNDAMENTAL PARTICLES.

ALSO THERE ARE LARGE NUMBERS OF THE OTHER DIFFERENT CLASSES OF NON DISCOVERED BIO FRIENDLY FUNDAMENTAL PARTICLES, WHICH THESE PARTICLES, PARTICLE CLOUDS, AND INFORMATION IMAGE PARTICLE CLOUDS CONSTRUCT THE MOST OF THE EARTH'S ATOMS, NANO UNITS, ELECTRONS, AND NUCLEON PARTICLE COMPOUNDS CONSTRUCTIONS, THESE LATER UNKNOWN LARGE CLASSES OF FUNDAMENTAL PARTICLES NEEDS NEW DISCOVERIES.

THE EARTH'S SENSIBLE INDIGENOUS BIO FRIENDLY FUNDAMENTAL PARTICLE INFORMATION IMAGE PARTICLE – CLOUDS (F.P. – I.I. – P. cl.), WHICH CONSTRUCT CENTRAL INTELLIGENCE SYSTEM CENTERS OF PLANETS EARTH'S INTELLECTUAL SYSTEM CENTERS OF THE DIFFERENT EARTH'S ATOMS, ELECTRONS, NUCLEONS, AND OTHER NANO – UNITS CONSTRUCTING STRUCTURES, ALL ARE CONSTRUCTED FROM THE USE OF THE FOLLOWING KNOWN BIO FRIENDLY FUNDAMENTAL PARTICLE CLASSES:

1 - DIFFERENT LIGHT FUNDAMENTAL PARTICLE CLASSES PARTICLE CLOUDS, AND Y - FP – I.I. – P. Cl.

2 – DIFFERENT SONIC FUNDAMENTAL PARTICLE CLASSES SONIC PARTICLE CLOUDS AND S- FP – I.I. – P. Cl.

3 – THE BIO FRIENDLY ELECTRIC FUNDAMENTAL PARTICLES, AND BIO FRIENDLY THERMAL FUNDAMENTAL PARTICLES, AND THEIR RELATED PARTICLE CLOUDS, SUCH AS ELECTRIC FUNDAMENTAL PARTICLES INFORMATIONS AND IMAGES PARTICLE CLOUDS (E. – FP – I.I. –P. Cl.), THERMAL FUNDAMENTAL PARTICLE INFORMATION IMAGE PARTICLE CLOUDS (T. – FP - -I.I. – P. Cl.), ETC. THESE ARE THE MOST COMMON FP – I.I. – P. Cl. THAT, THEY CONSTRUCT PARTICLE INTELLIGENCE SYSTEM CENTERS (PIS) OF THE DIFFERENT INTELLECTUAL ATOMS, ELECTRONS, NUCLEONS, AND OTHER INTELLECTUAL NANO- UNITS CIS PARTICLE COMPOUND CONSTRUCTIONS.

THE FUNDAMENTAL PARTICLE INFORMATION IMAGE PARTICLE CLOUD COMPOUND NEO - GENESIS (F.P. – I.I. – P.cl. – COMP. NEO - GENESIS) INSIDE INDIGENOUS ELECTRONS, NUCLEONS, ATOMS, NANO –UNITS CENTRAL INTELLIGENCE SYSTEM CENTERS (CIS), WHICH TAKE PLACE IN ALL PLANETS ACROSS THE UNIVERSE, UNDER A. S. I. FP. Mol. C.I.C., ALL ARE UNIVERSAL COMMON PHENOMENONS, AND ALL TAKE PLACE UNDER ORDINARY PHYSICS, CHEMISTRY, AND BIOLOGICAL LAWS AND RULES, WHICH WE SEE THOSE LAWS AND RULES IN PLANET EARTHS AS WELL, WITH NO DIFFERENCES.

IN THE DIFFERENT UNIVERSAL PLANETS, ALMOST THE INDIGENOUS FUNDAMENTAL PARTICLES ARE PRODUCERS OF THE PARTICLE CLOUDS, AND FUNDAMENTAL PARTICLES INFORMATION IMAGE PARTICLE CLOUDS (FP – I.I. – P. Cl.), THESE PLANET - NATIVE PARTICLES ALL ARE PLANET – SPECIFIC, ADDITIONALLY THE FUNDAMENTAL PARTICLES IN DIFFERENT PLANETS IN UNIVERSE, ARE DIFFERENT FROM EACH OTHERS.

THE UNIVERSAL EXISTING SUBJECTS, AND EXISTING ENVIRONMENTAL DIFFERENT THINGS, AND CONSTRUCTED INFORMATION IMAGE PARTICLE CLOUD COPIES FROM THESE SUBJECTS, ALL ARE DIFFERENT KIND FROM ONE PLANET TO THE OTHER. THE FUNDAMENTAL PARTICLES, PARTICLE CLOUDS, AND CONSTRUCTED PARTICLE COMPOUNDS FROM THESE PARTICLE CLOUDS, ALL DIFFERE FROM ONE PLANET TO OTHER.

THE FUNDAMENTAL PARTICLES CLOUDS, WHO ARE INDIGENOUS FOR GIVEN LOCATION PLANETS, AND THESE PARTICLE CLOUDS IN DIFFERENT UNIVERSAL LOCATIONS, AND THEIR PRODUCED FUNDAMENTAL PARTICLES INFORMATION AND IMAGE PARTICLE CLOUDS COPIES, AND PRODUCED PARTICLE COMPOUNDS AND FUNDAMENTAL PARTICLE INFORMATION AND IMAGE PARTICLE COMPOUNDS, ALL FROM ONE PLANET TO OTHER ARE REMARKABLY DIFFERENT.

THESE DIFFERENCES IN PARTICLE CLOUDS CONSTRUCTIONS, AND PARTICLE CLOUD COMPOUND CONSTRUCTIONS CAUSE PRODUCTIONS OF THE DIFFERENT STRUCTURE ELECTRONS, NUCLEONS, NANO – UNITS, AND ATOM CONSTRUCTIONS FROM ONE PLANET TO OTHER, UNLESS AT EXTREME RARE PROBABILITIES SITUATIONS, ONE MAY ENCOUNTER TWIN PLANETS, WHICH THEIR ENTIRE FUNDAMENTAL PARTICLE POPULATIONS AND THEIR CONSTRUCTED NANO – UNITS TO BE IDNTICAL.

THE CONSTRUCTED ELECTRONS, NUCLEONS, ATOMS, NANO – UNITS FROM INDIGENOUS DIFFERENT KIND FUNDAMENTAL PARTICLES, PARTICLE CLOUDS, AND FUNDAMENTAL PARTICLE INFORMATIONS AND IMAGE PARTICLE CLOUDS FROM ONE PLANET TO OTHER, ARE IMMENSELY DIFFERENT FROM EACH OTHER, AND PRODUCE IMMENSELY DIFFERENT KIND ELECTRONS, NUCLEONS, ATOMS, AND SUBJECT CONSTRUCTIONS.

THE BIO LETHAL PARTICLES SUCH AS; THE X-PARTICLE CLOUDS SUBJECT COPIES, THE GAMMA-PARTICLES INFORMATION AND IMAGE PARTICLE CLOUDS COPIES, AND OTHER BIO HOSTILE THERMAL, ELECTRIC, SONIC OR OTHER PARTICLES AND PARTICLE CLOUDS, FROM THE DIFFERENT PLANETARY AREAS SUBJECTS AND EXISTING THINGS ALL ARE ANOTHER CLASSES, THAT PRODUCE THEIR PARTICLE SPECIFIC FUNDAMENTAL PARTICLE CLOUDS, AND OBEY THE SAME EXACT RULES AND LAWS OF THE PARTICLES AND PARTICLE CLOUDS, NANO UNITS CONSTRUCTIONS PHYSICAL, CHEMICAL LAWS, AS WE HAVE IN PLANET EARTH.

IN DIFFERENT PLANETS INCLUDING THE PLANET EARTH, THE INDIGENOUS ATOMS ALSO CONSTRUCT AND PRODUCE NANO-UNIT - CLOUDS FROM OUTSIDE EXISTING THINGS AND SUBJECTS, ATOM INFORMATION IMAGE CLOUDS ALSO ARE ENVIRONMENTAL SUBJECTS COPIES MADE BY ATOM – CLOUDS.

PHENOMENON OF PRODUCTION OF INTELLECTUAL ATOMS

STORAGE OF PARTICLE CLOUDS

RI – GENESIS PHENOMENON INSIDE ATOMS

INSIDE ELECTRONS, NEUTRONS, PROTONS, ATOMS, AND ANOTHER NANO UNITS

PHENOMENON OF F.P. – I.I. – P. cl. _ COMP. – GENESIS,

GENESIS OF FACTUAL OR REAL INTELLIGENCE (RI) INSIDE ELECTRON, NEUTRON, AND PROTON

THE DIFFERENT BIO FRIENDLY S. Y. E. T. - FP – I.I. – P. Cl. OR CLOUDS, SUCH AS SONIC FUNDAMENTAL PARTICLE CLOUDS, LIGHT PARTILE CLOUDS, BIO FRIENDLY ELECTRIC FUNDAMENTAL PARTICLE CLOUDS AND THERMAL PARTICLE CLOUDS, FOLLOWING EMISSION FROM A CLOUD – PROVIDER SUBJECT, AIRBORN OR THROUGH INTERNAL P.A.S. PARTICLE TRANSIT - ROUTES, AND P.C.S. CURRENTS ENTER INTO INTERNAL ELECTRONS - NUCLEONS SUBSYSTEM UNITS CHEMICAL LAB.S OF THE RECIPIENT ATOMS THROUGH THE INTERNAL ATOM PARTICLE CIRCULATION SYSTEMS (PCS).

THE EXOGENOUS S. Y. T. E. – FP – I.I. – P. Cl. UNDER REGENERATIVE A. S. I. FP. Mol. C.I.C. INSIDE THE ELECTRONS, NUCLEONS SUBSYSTEM UNITS CHEMICAL LAB.S COMBINE WITH INTER - ELECTRON - NUCLEON PARTICLE COMPOUNDS, AND PRODUCE INTERNAL ELECTRON NUCLEON FUNDAMENTAL PARTICLE INFORMATION IMAGE MOLECULAR PARTICLE CLOUD – COMPOUND CONSTRUCTIONS. THIS IS PHENOMENON OF NEO – PARTICLE CLOUD COMPOUND GENESIS. WHICH THESE MOLECULAR PARTICLE CLOUDS CONSTRUCT NEW CONSTRUCTION INTELLECTUAL ELECTRONS, AND NUCLEONS.

THESE NEWLY CREATED INFORMATION IMAGE PARTICLE CLOUD COMPOUNDS INSIDE ELECTRONS, NUCLEONS CREATE PRODUCTIONS OF NEW INTELLECTUAL INFORMATIONS IMAGES INSIDE DIFFERENT NANO-UNITS STRUCTURES, WHICH THEY HAVE INFORMATIONS AND IMAGES OF EXOGENOUS ENVIRONMENTAL SUBJECTS INFORMATIONS AND IMAGES, THIS IS ALSO NEO-GENESIS OF INTELLECTUAL ELECTRONS, NEUTRONS, PROTONS AND ATOMS AND OTHER NANO UNITS, UNDER REGENERATIVE A. S. I. N.-U. C. I. C.

THIS IS PHENOMENON OF CREATIONS OF PARTICLE INTELLIGENCE SYSTEM CENTERS OF NANO UNIT CONSTRUCTIONS. THESE ARE PHENOMENONS OF NEO- PARTICLE COMPOUND GENESIS, PHENOMENON OF THE NEO- ELECTRON GENESIS, NEO- NUCLEON GENESIS, NEO NANO – UNITS GENESIS, AND CONSTRUCTIONS OF PARTICLE INTELLIGENCE SYSTEMS CENTERS FOR DIFFERENT NANO – UNITS. THROUGH CONSTRUCTIONS OF THE NEW INTELLECTUAL FUNDAMENTAL PARTICLE INFORMATION IMAGE PARTICLE CLOUD COMPOUNDS (F.P. – I.I. – P.cl. _ COMPOUNDS), WHICH THESE CENTERS POSSESS ENVIRONEMENTAL OUTSIDE INCIDENTS IMAGES AND INFORMATIONS IN THE PARTICLE COMPOUND FORMS ABOUT DIFFERENT EXTERNAL SUBJECTS AND MATTERS. WHICH ALL STORED INSIDE THE INTELLECTUAL ATOMS, ELECTRONS, NUCLEONS, CENTRAL INTELLIGENCE SYSTEM CENTERS.

THIS IS INTELLIGENCE CAPABILITY - CREATIONS THROUGH STORING INFORMATIONS AND IMAGES IN PARTICLE CLOUD FORMS INSIDE THE ELECTRONS, NUCLEONS, ATOMS CENTRAL INTELLIGENCE SYSTEM CENTERS, UNDER REGENERATIVE A. S. I. FP. Mol. C.I.C.

THE CENTRAL INTELLIGENCE SYSTEM CENTERS (CIS) OF ELECTRONS, NUCLEONS, ATOMS STORE OUT SIDE ENVIRONMENTAL INFORMATION IMAGE PARTICLE CLOUDS IN FORMS OF PARTICLE CLOUD- COMPOUNDS CONSTRUCTIONS INSIDE CENTRAL INTELLIGENCE CENTERS.

THIS PHENOMENON IS STORING AND RECORDING OF OUTSIDE ENVIRONMENTAL KNOWLEDGE INSIDE ELECTRONS NUCLEONS IN THE FORMS OF F.P. – I.I.- P. cl. _COMPOUND MOLECULAR CONSTRUCTIONS.

RETRIEVAL OF PARTICLE CLOUDS

THE MEMORIZATION PHENOMENON

2 - FUNDAMENTAL PARTICLE INFORMATION IMAGE PARTICLE CLOUDS (F.P. – I.I. – P. cl.) RETRIEVAL PHENOMENON

PHENOMENON OF PSYCHE GENESIS, THOUGHT CURRENT GENESIS, AND REMEMBERING:

BREAKING DOWN OF LARGE MOLECULAR PARTICLE CLOUD STRUCTURES SUCH AS F.P. – I.I. – P. cl. _ COMPOUNDS INSIDE CNS ELECTRONS NUCLEONS, TAKE PLACE UNDER DEGENERATIVE A. S. I. F.P. Mol. C.I.C., AND UNDER THIS PROCESS THE COMBINED INFORMATION-IMAGE PARTICLE CLOUDS COMPOUNDS BREAK DOWN INTO CONSTRUCTING SMALLER MOLECULAR STRUCTURES, AND CAUSE RELEASE OF (F.P. – I.I.- P.cl.) PARTICLE CLOUDS, A FREE INTERACTING PARTICLE – CLOUDS CURRENTS INSIDE THE CNS ATOMS PARTICLE CLOUD CIRCULATION SYSTEMS.

UNDER DEGENERATIVE A. S. I. F.P. – P.cl. Mol. C.I.C., WHICH IT IS REVERSE DIRECTIONS OF ORIGINAL REGENERATIV AUTONOMOUS SEQUENTIAL CHEMICAL INTERACTIONS CYCLES, CAUSE BREAK DOWN OF LARGE PARTICLE CLOUD COMPOUNDS INTO SMALLER FREE MOLECULES, AND THIS PHENOMENON IS PARTICLE CLOUD RETRIEVAL PROCESS AND CAUSE RELEASE OF ENVIRONMENTAL SUBJECTS F.P.- I. I. - P. cl. FREE, AND THESE PARTICLE CLOUDS INTER INTO CNS PARTICLE CIRCULATION SYSTEMS AS FREE INTERACTING F.P. –I.I. - P. cl. (FREED KNOWLEDGE OF EVENTS, IN PARTICLE CLOUD FORMS), THIS IS RETRIEVAL PHENOMENON OF PARTICLE CLOUDS INTO FREE PARTICLE CLOUDS FORMS, WHICH UNDER COMPUTER TECHNOLOGY CAN BE WATCHED ON TV SCREENS.

THE FREE INFORMATION IMAGE PARTICLE CLOUD CURRENTS (F.P. – I.I.- P. cl. CURRENTS) THROUGH PARTICLE CIRCULATION SYSTEMS TRAVEL BETWEEN DIFFERENT BRAIN CENTERS AND INTERACT WITH PARTICLE CLOUDS OF DIFFERENT CNS CENTERS ELECTRONS NUCLEONS, THESE PARTICLE CLOUDS CURRENTS AND INFORMATION IMAGE INTERACTIONS WITH ALL CNS ELECTRONS NUCLEONS PARTICLE COMPOUNDS, BIOLOGICALLY SENSE AS THOUGHT CURRENTS AND PSYCHE, THIS IS PHENOMENON OF THOUGHT CURRENT GENESIS AND PHENOMENON OF PSYCHE GENESIS.

IN THE PHENOMENON OF RE- MEMORIZATION OF PAST EVENTS, WHEN THE PAST EVENTS RECORD IN F.P. – I. I. – P.cl. COMPOUND FORMS, INSIDE THE CNS ELECTRONS NUCLEONS, UNDER RETRIEVAL PROCESS, WHEN LARGE MOLECULAR INFORMATION IMAGE PARTICLE CLOUD COMPOUNDS BREAK DOWN INTO SMALLER MOLECULAR STRUCTURES UNDER DEGENRATIVE AUTONOMOUS CHEMICAL INATERACTIONS ONE BY ONE.

DURING RETRIEVAL PHASE, THE RELEASED PARTICLE INFORMATION AND IMAGE CLOUDS ARE EXACT VIRTUAL COPIES OF THE ORIGINAL SUBJECTS, AND SCENES, DURING RETRIEVAL OF PARTICLE CLOUDS THE EVENT OF THE PAST RE-APPEARING AGAIN INSIDE CNS ELECTRONS NUCLEONS EXACTLY THE SAME AS ORIGINAL SUBJECTS PLAYIN SCENES AND EVENTS SIMILARLY, THIS RETRIEVAL PHENOMENON AS KNOWN AS CNS MEMORIZATION AND REMEMBERING PHENOMENON, IN REALITY THE VIRTUAL PARTICLE CLOUDS INSIDE CNS ELECTRONS NUCLEONS, LOOKS EXACTLY THE SAME AS ORIGINAL SCENES- SUBJECTS AND EVENTS, AND BOTH ORIGINAL SUBJECTS AND VITUAL CLOUDS ARE IDENTICAL, LOOKS SIMILAR, THAT IS THE REASONS THE THOUGHTS,PSYCHE, THOUGHT CURRENTS WHICH ARE PARTICLE CLOUD INTERACTIONS AND CURRENTS INSIDE THE CNS ELECTRONS NUCLEONS, ALL LOOKS TWINS EVENTS OF THE ORIGINAL SUBJECTS AND SCENES.

INTERNAL ATOM, INTER-ELECTRON, INTER-NUCLEON PARTICLE CIRCULATION SYSTEMS (PCS):

THE ONGOING PARTICLE CURRENTS AND PARTICLE CLOUDS CIRCULATIONS SYSTEMS IN ATOM'S PERIPHERON, NUCLEUS, AND PERIPHERAL ATOM SPACE (PAS) ARE AS FOLLOWING:

1 - THE PERIPHERON'S PARTICLE CIRCULATION SYSTEMS (P-PCS)

==

THE PERIPHERON HAVE TWO PARALLEL PARTICLE CIRCULATION SYSTEMS AND PARTICLE CURRENTS,

THE FIRST PARTICLE CURRENTS CIRCULATE FROM PERIPHERAL ATOM SPACE TOWARD THE ATOM'S CENTER, THIS CENTRIPETAL P.P.C.S. BRINGS NEEDED NECESSARY PARTICLE CLOUDS AND NEEDED NANO-UNITS, PARTICLES FROM OUTSIDE ATOM AND PERIPHERAL ATOM SPACE INTO ATOM, TO BE USED FOR NEEDED NECESSARY CHEMICAL, PHYSICAL, BIOLOGICAL FUNCTIONS, A.S. I. F.P. Mol. C.I.C. AND PARTICLE CONSTRUCTIONS OF THE ELECTRONS, AND NEEDED PARTICLES AND PARTICLE CLOUDS TO ACHIEVE INTERNAL ELECTRON NANO-TASKS,

THE CENTRIPETAL P.P.C.S. BRING NEEDED PARTICLES AND PARTICLE CLOUDS FROM OUTSIDE FOR CONSTUCTIONS OF INTERNAL ELECTRON NUCLEON PARTICLE COMPOUND COSTRUCTIONS, AND CONSTRUCT ELECTRONS NUCLEONS THROUGH USE OF THESE INCOMING PARTICLES AND PARTICLE CLOUDS,

THE SECOND PPCS, RUNS AND CIRCULATE CENTRIFUGALLY PARALLEL BUT IN OPPOSITE DIRECTIONS OF THE FIRST PARTICLE CIRCULATION SYSTEM, THE SECOND PPCS CURRENTS FLOW AND CIRCULATE FROM ATOM'S CENTER OR NUCLEUS TOWARD THE PERIPHERAL ATOM IN CENTRIFUGAL DIRECTION PARTICLE CLOUD CURRENTS AND PARTICLE CIRCULATION SYSTEMS,

THE SECOND PPCS TRANSPORT AND CARRY OUT UN- NEEDED PARTICLE BY PRODUCTS, ALSO CARRY OUT CONSTRUCTED AND PRODUCED PARTICLE CLOUDS AND PARTICLE COMPOUNDS WHICH HAVE BEEN PRODUCED BY ELECTRONS NUCLEONS FOR THE USE OF ANOTHER EXTERNAL NANO-UNITS NEEDS, FOR USE OF ANOTHER AT RECIPIENT NANO-UNITS CONSTRUCTIONS.

THE PERIPHERONS PPCS COMPOSED FROM TWO DISTINCT CIRCULATION SYSTEMS, ONE PPCS CIRCULATE IN PERIPHERON, AND THE OTHER PPCS IS INTERNAL ELECTRONS PARTICLE CIRCULATION SYSTEMS,

THE PERIPHERON'S PARTICLE CIRCULATION CURRENTS ALSO CROSS PERIPHERON LINES INTO NUCLEUS AND JOIN TO NUCLEI PARTICLE CIRCULATION SYSTEMS, THE PERIPHERON PARTICLE CIRCULATION SYSTEMS FURTHER MORE IN PERIPHERON BRANCH INTO DIFFERENT SMALL SIZE INTERNAL ELECTRONS BRANCHES AND CONNECT TO INTERNAL ELECTRONS PARTICLE CIRCULATION SYSTEMS,

2 - THE NUCLEAR PARTICLE CIRCULATION SYSTEMS(N-PCS)

AND PARTICLE CLOUD CURRENTS

THE NUCLEUS PARTICLE CIRCULATION SYSTEMS INCLUDE, INTERNAL PROTONS CIRCULATION SYSTEM, INTERNAL NEUTRONS PARTICLE CIRCULATION SYSTEMS, AND INTERNAL NUCLEUS PARTICLE CIRCULATION SYSTEMS, THE INTERNAL NUCEUS PARTICLE CIRCULATION SYSTEMS PARTICLE CURRENTS SIMILAR TO P.P.C.S., FLOWS AT TWO OPPOSITE DIRECTIONS PARTICLE CIRCULATION SYSTEMS CURRENTS.

THE FIRST NUCLEAR PARTICLE CIRCULATION SYSTEMS (N.P.C.S.) CURRENTS FLOW CENTRIPETALLY FROM PERIPHERON TOWARD THE NUCLEAR CENTER DIRECTIONS, AND BRINGS NEEDED PARTICLES AND PARTICLE CLOUDS INTO NUCLEUS, PROTONS, NEUTRONS, FOR NEUTRON-GENESIS AND PROTON-GENESIS AND CONSTRUCTIONS OF THE NUCLEON PARTICLE COMPOUND CONSTRUCTIONS, THE INCOMING PARTICLES AND PARTICLE CLOUDS ARE USED TO CONSTRUCT INTERNAL NUCLEONS PARTICLE COMPOUNDS AS WELL ARE USED ACHIEVE NUCLEONS PHYSICAL CHEMICAL BIOLOGICAL NANO-TASKS.

THE SECOND N. P.C.S. AND PARTICLE CURRENTS FLOW PARALLEL TO THE FIRST NUCLEAR PARTICLE CIRCULATION SYSTEMS CENTRIFUGALLY, AND CARRY OUT UN-NEEDED PARTICLE BY PRODUCTS, AND PRODUCED PARTICLE MOLECULES FOR USE OF NANO-UNIT PARTICLE COMPOUND CONSTRUCTIONS AND FOR USE OF OTHER NANO-TASKS.

3 – THE PARTICLE CIRCULATION SYSTEMS AND PARTICLE

CURRENTS OF PERIPHERAL ATOM SPACE (PAS-PCS)

====================================

THE PERIPHERAL ATOM SPACE PARTICLE CIRCULATION SYSTEM'S CURRENTS, CIRCULATE BETWEEN DIFFERENT ATOMS AND IN THE PERIPHERY OF ATOMS, THE ATTRACTIVE GRAVITON FORCES OF ELECTRONS, NUCELONS ATTRACT THEIR NEEDED PARTICLES FROM THEIR SPACE (PAS), THE PARTICLE CLOUDS AND INFORMATION IMAGE PARTICLE CLOUDS FROM PERIPHERAL ATOM SPACE (P.A.S.) UNDER ATTRACTION GRAVITON FORCES INTER INTO PERIPHERON PARTICLE CIRCULATION SYSTEMS (PPCS), AND THROUGH THE PERIPHERONS PARTICLE CIRCULATION SYSTEMS (PPCS) THE ELECTRONS NEEDED PARTICLES, PARTICLE CLOUDS CIRCULATE INTO DIFFERENT ELECTRONS IN THE PERIPHERON, THEREAFTER THROUGH NUCLEAR PARTICLE CIRCULATION SYSTEMS (N.P.C.S.) THE INCOMING PARTICLES , F.P.- I.I.- P. cl., AND OTHER PARTICLE CLOUDS CIRCULATE INSIDE NUCLEUS AND BRANCH TO DIFFERENT EXISTING PROTONS AND NEUTRONS SUBSYSTEM –UNITS CONSTRUCTIONS, AND COMBINE WITH THEIR PARTICLE COMPOUNDS OR USED FOR ACHIEVING DIFFERENT NANO- FUNCTIONS INSIDE THE DIFFERENT ELECTRONS NUCLEONS.

THE PPCS, AND NPCS PROVIDE NEEDED PARTICLE CLOUDS, FUNDAMENTAL PARTICLES, F.P. – I. I. - P.cl. INTO DIFFERENT ELECTRONS AND NUCLEONS SUBSYSTEM - UNITS CHEMICAL LAB.S, THERE DIFFERENT ELECTRONS NUCLEONS PARTICLE COMPOUNDS UNDER A. S. I. F.P. Mol. C.I.C. COMBINE WITH INCOMING PARTICLE CLOUDS AND PRODUCE NEW CONSTRUCTIONS PARTICLE COMPOUNDS, WHICH THESE ARE THE PARTICLE COMPOUNDS CONSTRUCTING THE DIFFERENT ELECTRONS NUCLEONS PARTICLE STRUCTURES, THID IS PROCESS OF ELECTRONS NEO-GENESIS AND NUCLEONS NEO-GENESIS PHENOMENONS.

Communicating of Atoms with each other

STORAGE OF ENVIRONMENTAL KNOWLEDGE OF UNIVERSE IN INFORMATION IMAGE PARTICLE CLOUDS FORMS, INSIDE ELECTRONS NUCLEONS. EXCHANGE OF THESE F.P.-I.I.- P. cl. BETWEEN DIFFERENT ATOMS:

IN EARTH THE BIOFRIENDLY LIGHT, SONIC, ELECTRIC, THERMAL FUNDAMENTAL PARTICLE INFORMATION IMAGE PARTICLE CLOUDS (Y. S. E. T. – F.P. – I.I. – p.cl.) ARE THE MOST COMMON TYPES PARTICLE CLOUDS, TRANSIT AND EXCHANGE INTELLIGENCE IN FORMS OF PARTICLE CLOUDS INFORMATIONS AND IMAGES, BETWEEN DONOR AND RECIPIENTS ELECTRONS NUCLEON, AND INTERACT WITH EACH OTHER UNDER A. S. I F.P. Mol. C.I.C. IN THE EARTH, AND SIMILARLY IN ENTIRE UNIVERSE BETWEEN THE ATOMS.

THIS PHENOMENON IS EXCHANGES OF EXISTING INTELLIGENCES AND KNOWLEDGES OF PLANETS AND UNIVERSES, IN FUNDAMENTAL PARTICLE CLOUDS INFORMATIONS IMAGES FORMS, AND THIS PHENOMENON IS COMMUNICATIONS OF EXISTING PLANETARY AND UNIVERSAL INTELLIGENCES AND KNOWLEDGES, IN THE FORMS OF Y. S. T. E. – F.P. – I.I. – P. cl., BETWEEN THE MOST OF EXISTING ELECTRONS, NUCLEONS.

IN PERFORMING THESE NANO- TASKS, SOME ELECTRONS, NUCLEONS, AND NANO- UNITS ARE MUCH INTELLIGENCE, AND CAPABLE THAN THE OTHERS TO PERFORM THESE INTER NANO-UNIT COMMUNICATIONS AND STORAGES BETTER THAN OTHERS. EVEN SOME OF THE NANO – UNITS ARE ENTIRELY ILLITERATE IN THESE ESSENTIAL UNIVERSAL TASKS, AND CAN NOT COMMUNICATE AND STORE ANY INFORMATION INAGES CLOUDS AT ALL.

COMBINING OR STORING S.Y.-F.P.-I.I.-P.cl.(PARTICLE CLOUDS OF OUTSIDE), WITH INSIDE ELECTRONS NUCLEONS:

RECORDING AND GENESIS OF OUTSIDE WORLD'S INFORMATION IMAGE ENCYCLOPEDIA INSIDE CNS ATOMS

PHENOMENON OF PSYCHE-GENESIS AND THOUGHT GENESIS:

GENESIS OF FACTUAL INTELLIGENCE (RI) INSIDE ELECTRONS, NUCLEONS, AND ATOM

EXISTING SUBJECTS AT EARTH, EMIT PARTICLE CLOUDS COPIES FROM THEMSELVES, WHICH THESE PARTICLE CLOUDS COPIES HAS BEEN MADE MOSTLY FROM LIGHT AND SONIC FUNDAMENTAL PARTICLE CLOUDS (S. Y. – F.P. – I.I. – P. cl.) IN PLANET EARTH, THESE PARTICLE CONSTRUCTED CLOUDS ARE IN EXACT FORMS OF ORIGINAL SUBJECTS, BUT THESE SUBJECT'S - COPY CLOUDS ALL MADE FROM FUNDAMENTAL PARTICLES IN CLOUD FORMS. THESE PARTICLE CLOUDS COPIES OF EXISTING SUBJECTS ARE CAPABLE TO MOVE, SPEAK, PERFORM, WALK, STUDY AND DO ANY THINGS THAT ORIGINAL SUBJECTS IS CAPABLE TO DO EXACTLY THE SAME, AS ORIGINAL SUBJECT'S CAPABILITIES.

THE CLOUDS ARE INFORMATIONS, IMAGES, PERFORMANCES IN CLOUD FORMS, MADE FROM DIFFERENT PARTICLES. THE CLOUDS ALL MADE FROM FUNDAMENTAL PARTICLES, WHEN THESE CLOUDS (S. Y. –F.P. – I.I. – P. cl.) RECORDS AND STORES INSIDE CNS ELECTRONS AND NUCLEONS IN PARTICLE CLOUD COMPOUND FORMS AND CONSTRUCT THE CNS ELECTRONS NUCLEONS PARTICLECOMPOUND CONSTRUCTIONS, THESE STORED A. S. Y.- F.P. – I.I. – P. cl. ARE EXACTLY SIMILAR TO ORIGINAL SUBJECTS PERFORMANECES INSIDE THE ELECTRONS AND NUCLEONS OF CNS. CLOUDS PERFOMANCES INSIDE CNS ELECTRONS NUCLEONS ARE EXACTLY THE SAME AS ORIGINAL SUBJECTS PERFORMING INSIDE BRAIN. THIS IS THE WHAT THE ANIMALS FEEL INSIDE THEIR BRAINS WHEN THEY ARE

THINKING. (WHAT THEY ARE SEEING INSIDE THEIR BRAINS, ALL ARE PARTICLE CLOUD PERFORMANCES, AND THERE IS NO REAL SUBJECT INSIDE ELECTRONS AND NUCELONS OF THE CNS).

IN ALL PLANETS, THE ALL FUNDAMENTAL PARTICLE INFORMATION IMAGE CLOUDS, MOSTLY ARE CONSTRUCTED FROM PLANET'S NATIVE FUNDAMENTAL PARTICLES, FROM THE FUNDAMENTAL PARTICLES THAT THOSE ARE SPECIFIC AND BELONG AS NATIVE PARTICLES FOR THOSE PLANETS.

PARTICLE CLOUD DONERS

AT EARTH MOST COMMON INFORMATION AND IMAGES PARTICLE CLOUDS ARE MADE FROM BIOFRIENDLY: LIGHT PARTICLES, SONIC PARTICLES, ELECTRIC PARTICLES, THERMAL FUNDAMENTAL PARTICLE CLOUDS (Y. S. E. T. – F.P. – I.I. – p. cl.), THE MOST OF SUBJECTS AND EXISTING THINGS OF PLANETH EARTH ARE CAPABLE TO PRODUCE AND EMIT Y. S. T. E. – F.P. – I.I. – p.cl. INTO AIR OR TRANSIT THROUGH OTHER FUNDAMENTAL PARTICLE TRANSITION ROUTES, THE PARTICLE CLOUD DONERS (F.P. – I.I. – p. cl. – DONERS) ARE THOSE SUBJECTS THAT, THEY EMIT AND PRODUCE FROM THEMSELVES F.P. – I.I.- P. cl. - COPY INTO SPACE OR AIR.

THE EXISTING LIVING THINGS, SENSARY ORGANS ELECTRONS-NUCLEONS POSSESS CAPABILITIES UNDER ATTRACTIVE GRAVITON FORCE TO ATTRACT PARTICLE CLOUDS FROM AIR, AND CAPTURE THE PARTICLE CLOUDS FROM AIR AS RECIPIENTS OF INFORMATION IMAGES PARTICLE CLOUDS FROM ANY SOURCE. THIS PHENOMENON IS COMMUNICATIONS BETWEEN DIFFERENT INTELLIGENCE SYSTEMS, THE ATTRACTED PARTICLE-CLOUDS THROUGH NANO NEURAL FIBERS TRANSMIT INTO CNS ELECTRONS NUCLEONS, COMBINE WITH CNS SENSARY CENTERS PARTICLE COMPOUNDS, AND CONSTRUCT THE INTER ELECTRON NUCLEONS PARTICLE COMPOUND CONSTRUCTIONS.

THIS IS PHENOMENON OF INTELLIGENCE EXCHAGE BETWEEN ATOMS ELECTRONS NUCLEONS, THE A. S. I. F.P. Mol. C.I.C. POSSESS EXTREME RESPONSIBILITIES OF RECORDING, INTERACTIONS, AND RETRIEVAL PHENOMENON OF THE INFORMATIONS AND IMAGES PARTICLE CLOUDS INSIDE ATOMS, OR TO OUTSIDE ELECTRONS, NUCLEONS.

PARTICLE CLOUD (Y. S. E. T. – F.P. – I.I. – p. cl.)- CURRENTS CIRCULATIONS SYSTEMS

BETWEEN PARTICLE DONERS AND RECIPIENTS

THE PRODUCED AND EMITTED PARTICLE CLOUDS (Y. S. E. T. – F.P. –I.I. – P. cl.) FROM DONER'S SUBJECTS TRAVEL AIRBORN IN AIR AT ANY DIRECTION, UNTIL THE RECIPIENT'S SUBJECTS SENSARY ELECTRONS-NUCEONS, THROUGH THEIR ATTRACTIVE FORCES OF GRAVITON ATTRACT PARTICLE CLOUDS INTO INSIDE RECIPIENT ELECTRONS-NUCLEONS SUBSYSTEM-UNITS CHEMICAL LAB.S, THE INCOMING EXOGENOUS INFORMATION IMAGE PARTICLE CLOUDS COMBINE WITH PARTICLE COMPOUNDS OF RECIPIENT ELECTRONS NUCLEON, AND PRODUCE INFORMATION-IMAGE PARTICLE CLOUD - COMPOUNDS AND BECOME PART OF CONSTRUCTIONS OF RECIPIENT ATOMS. (S. Y. T. E. – F.P. – I.I. – p. cl.- COMP.) THIS IS PHENOMENON OF STORAGE OF ENVIRONMENTAL SUBJECTS INFORMATION IMAGES INSIDE RECIPIENT ATOMS, IN PARTICLE CLOUD COMPOUND FORMS,

STORING SCIENCE AND KNOWLEDGE INSIDE ELECTRONS-NUCLEONS

INTELLIGENCE – GENESIS OF ELECTRONS, NUCLEONS

AIRBORN PARTICLE CLOUD CIRCULATION SYSTEMS BETWEEN ANIMALS AND OTHER EXISTING THINGS

THROUGH AIRBORN PARTICLE –CLOUDS (S. Y. – F.P. – I.I. – P. cl.) CIRCULATION SYSTEMS, BETWEEN DIFFERENT INDIVIDUALS, EITHER THEY ARE PARTICLE CLOUD EMITTERS, AND OTHER INDIVIDUALS ARE PARTICLE CLOUD RECEIVERS, THROUGH AIRBORN CLOSED CYCLE PARTICLE CLOUD CIRCULATION SYSTEMS BETWEEN DIFFERENT CNS ELECTRONS AND NUCLEONS DURING A COMMUNICATION, SOME INDIVIDUALS EMIT PARTICLE CLOUDS, AND OTHERS RECEIVES PARTICLE CLOUDS, THESE F.P. – I.I.- P. cl. CIRCULATE INSIDE A AIRBORN CLOSED CIRCUIT OF PARTICLE CLOUD CIRCULATION SYSTEMS BETWEEN INDIVIDUALS CNS ELECTRONS NUCLEONS.

THE DIFFERENT INDIVIDUALS THROUGH EXCHANGES OF S. Y. –I.I.- P. cl. BETWEEN EACH OTHER, THEY ANALIZE THE PARTICLE CLOUDS THROUGH A.S. I. F.P. Mol. C.I.C. INSIDE CNS ELECTRONS NUCLEONSAND CONSTRUCT THE PARTICLE CLOUD RESPONSES.

DIFFERENT SUBJECTS EMIT PARTICLE –CLOUDS – COPY OF SELF INTO SPACE, THE PRODUCED SONIC- LIGHT FUNDAMENTAL PARTICLE INFORMATION IMAGE PARTICLE CLOUDS FROM A GIVEN SUBJECT (S. Y. – I.I. - F.P. –P.cl.) TRAVEL AIRBORN OR THROUGH OTHER PARTICLE TRANSMISSION ROUTES FROM DONER SUBJECT TOWARD RECIPIENT ATOMS.

THE RECIPIENT ATOMS THROUGH THEIR ATTRACTIVE GRAVITON FORCES ATTRACT INCOMING S. Y. – F.P. – I.I. – P. cl. INTO P. A. S. AND THEREAFTER PARTICLE – CLOUDS FROM P.A.S INTER INTO INSIDE RECIPIENT ELECTRONS-NUCLEONS SUBSYSTEM –UNITS AND COMBINE WITH INTER ELECTRONS NUCLEONS DIFFERENT PARTICLE COMPOUNDS AND PRODUCE S. Y. – F.P. – I.I. – P. cl. _ COMPOUNDS AND BUILD RECIPIENTS ATOMS INTERNAL ELECTRON-NUCLEON PARTICLE COMPOUND CONSTRUCTIONS, AND STORE S.Y. – F.P. –I.I. – P. cl. INSIDE RECIPIENT ELECTRONS – NUCLEONS IN PARTICLE – CLOUD _ COMPOUND FORMS, THIS IS PHENOMENON OF STRORAGE OF DIFFERENT ENVIRONMENTAL SUBJECTS INFORMATION – IMAGE PARTICLE – CLOUDS INSIDE ATOMS, THIS PHENOMENON IS ALSO STORAGE OF INFORMATION-IMAGE OR KNOWLEDGE AND SCIENCE INSIDE ELECTRONS – NUCLEONS, WHICH UNDER DEGENERATIVE A. S. I. F.P. Mol. C.I.C. CAN BE RETRIEVED AS FREE PARTICLE-CLOUDS SCAPE TO OUT OF ELECTRONS –NUCLEONS INTO SPACE AGAIN.

PARTICLE-CLOUDS CAN TRANSIT THROUGH PERIPHERAL ATOM SPACE OF SOLIDS, LIQUIDS, PLASMA FROM ONE LOCATION TO OTHER THROUGH PRODUCED PARTILE-CURRENTS, ATOM-CLOUDS ARE DIFFERENT SUBJECTS, BUT FOLLOW THE SAME PARTICLE BASE RULES,

S. Y. - F. P. - I. I. – P.cl. FROM PERIPHERAL ATOM-SPACE (P.A.S.) INTER INTO PERIPHERON, TRAVEL THROUGH PERIPHERON-NUCLEUS FUNDAMENTAL-PARTICLE CIRCULATION SYSTEMS (P. C. S.) PARTICLE CURRENTS IN PERIPHERON AND NUCLEUS DIVID INTO SMALLER PATICLE CIRCULATION BRANCHES, PARTICLE CLOUDS FINALLY FROM SIDE BRANCHES, INTER INTO DIFFERENT ELECTRONS PROTONS AND NEUTRONS SUBSYSTEMS-UNITS CHEMICAL -LAB.S, AND COMBINE WITH DIFFERENT ELECTRON-NUCLEON PARTICLE COMPOUNDS AND PRODUCE NEW S. Y. –F.P. – I.I. – pl. _ COMPOUNDS CONSTRUCTING, CONSTRUCT THE NEW ELECTRON-NUCLEON PARTICLE COMPOUND CONSTRUCTIONS, THAT POSSESS SPECIFIC INFORMATIONS KNOWLEDGE IN REGARD TO SPECIFC OUTSIDE FACTS, INFORMATIONS IMAGES ARE STORED INTACT INSIDE ATOMS (PHENOMENON OF INTELLIGENCE ELECTRONS-NUCLEONS GENESIS), THIS IS RECORDING OF INFORMATIONS – IMAGES OF OUTSIDE FACTS INSIDE

THOUGHT CURRENT GENESIS AND PSYCHE – GENESIS PHENOMENONS

PHENOMENONS OF DECISION MAKING, MEMORIZATION AND ORDER ISSUING PHENOMENONS BY CNS CENTERS ELECTRONS –NUCLEONS:

IN LIVING THINGS INSIDE CNS ELECTRONS NUCLEONS THROUGH A. S. I. F.P. Mol. C.I.C. THE PRE-EXISTING CONSTRUCTED INDIGENOUS S. Y. –F.P. –I.I.- P. cl. _ COMPOUNDS COMBINE WITH INCOMING EXOGENOUS Y. S. – F.P. – I.I. – P. cl. AND PRODUCE DIFFERENT PARTICLE BY PRODUCTS, PARTICLE CLOUD COMPOUNDS, THE DIFFEREN PARTICLE CLOUDS CIRCULATE BETWEEN DIFFERENT CNS CENTERS THROUGH, CNS PARTICLE CIRCULATING SYSTEMS BETWEEN DIFFERENT CNS CENTERS ELECTRONS NUCLEONS, THE PARTICLE CLOUDS INTERACT WITH DIFFERENT OTHER CNS CENTERS INTER ELECTRON- NUCLEON PARTICLE CLOUD COMPOUNDS, ARE IN CONTINUEOUS COMPLEX A. S. I. F.P. Mol. C.I.C., ALL OF THESE PARTICLE CLOUD INTERACTIONS AND CIRCULARTION OF PARTICLE CLOUD CURRENTS INSIDE CNS. PARTICLE CIRCULATION SYSTEM, THE PHENOMENONS BIOLOGICALLY SENSE BY LIVING THINGS AS THOUGHT-CURRENTS AND PSYCHE PHENOMENON.

IN EARTH THE LIVING THINGS LIGHT- SOUND SENSER ORGANS ELECTRONS – NUCLEONS CAPTURE INCOMING EXOGENOUS Y. S. – F.P. – I.I. – P. cl. OF ENVIRONMENTAL EXISTING THINGS PARTICLE CLOUDS, THROUGH ATTRACTIVE GRAVITON FORCES THE PARTICLE CLOUDS INTER INTO ELECTRONS NUCLEONS CHEMICAL LAB.S, AND INCIDENT PARTICLE CLOUDS CHEMICALLY COMBINE WITH NTER ELECTRONS NUCLEON PARTICLE COMPOUNDS. PRODUCE PARTICLE CLOUD COMPOUNDS. AND THERE AFTER THROUGH NEURAL PARTICLE TRANSIT – ROUTES, AN PARTICLE CLOUD CIRCULATION SYSTEMS, THE EXOGENOUS INFORMATION IMAGE PARTICLE-CLOUDS CARRIED INT DIFFERENT CNS- CENTERS ELECTRONS- NUCLEONS SUBSYSTEM UNITS CHEMICAL LAB.S, AND COMBINED WITH PARTICLE COMPOUNDS, BUILD CNS SENSARY CENTERS ELECTRONS- NUCLEONS PARTICLE COMPOUND CONSTRUCTIONS, WHICH THE PARTICLE CLOUD COMPOUNDS ARE S. Y. –F.P. – I.I. – p. cl. –STORAGE SYSTEMS.

THE S. Y. - F.P. –I.I. – P.cl. ARE KNOWLEDGE OF OUTSIDE EVENT'S INFORMATION AND IMAGES IN FORMS OF PARTICLE CLOUDS, THAT ARE STORED IN PARTICLE COMPOUND FORMS, AT INSIDE CNS SENSARY CENTERS ELECTRONS NUCLEONS, THESE PARTICLE CLOUD STORAGES ARE ENVIRONMENTAL KNOWLEDGE AND SCIENCE STORING SYSTEMS, INSIDE CNS ELECTRONS AND NUCLEONS IN FORMS OF S. Y. – F.P. –I.I.- p.cl. _COMPOUNDS, WHICH THESE STORAGES ARE THE BUILDING BLOCKS OF LEARNING, SCHOOLING, TEACHING, EDUCATIONS, TRAININGS, ETC.

THROUGH RECORDING, STORING OF OUTSIDE WORLD S. Y. – F.P. – I.I. – p. cl. IN PARTICLE CLOUD- COMPOUNDS FORMS INSIDE SENSARY CNS CENTERS ELECTRONS-NUCLEONS, THESE KNOWLEDGE STOCK PILES AND INTELLIGENC STORAGE SYSTEMS PHENOMENONS OF OUTSIDE WORLD EVENTS INFORMATIONS AND IMAGES AT INSIDE CNS ELECTRONS AND NUCLEONS, ARE PHENOMENON OF ACCUMULATIONS OF SCIENCE AND KNOWLEDGE OF OUTSIDE INFORMATION, IN PARTICLE-CLOUD COMPOUND FORMS, AT INSIDE CNS ATOMS, THIS PHENOMENON IS BUILDING BLOCKS OF LEARNING, TRAINING, TEACHING, EDUCATIONS, AT HOMES, SCHOOLS, UNIVERSITY AS WELL AT ALL SOCIAL, RELIGIOUS, POLITICAL, AND EDUCATIONAL FIELDS.

THE A. S. I. F.P. Mol. C.I.C. BETWEEN THESE INDIGENOUS S. Y. – F.P. – I.I. – p. cl. _ COMPOUNDS WITH OTHER INCOMING EXOGENOUS INCOMING S. Y. – F.P. – I.I. – p. cl. INSIDE CNS ELECTRON- NUCLEON SUBSYSTEM – UNITS CHEMICAL LAB.S, AND INTERACTIONS OF ALL OF THESE WITH ENTIRE CNS DIFFERENT CENTER ELECTRONS – NUCLEON POPULATIONS, WITH EXPONENTIAL SPEEDS INTERACT WITH EACH OTHER, PRODUCE PARTICLE CLOUDS AND PARTICLE CLOUD CURRENTS, AND THEIR CHEMICAL INTERACTIONS WITH EACH OTHER CAUSE PRODUCTIONS OF NEW PARTICLE BY PRODUCTS, WHICH ALL OF THESE EVENTS BIOLOGICALLY SENSED AS PHENOMENON OF PSYCHE AND THOUGHT CURRENTS BY LIVING THINGS, THIS IS CAUSE OF PSYCHE-GENESIS, THOUGHT - GENESIS AND THOUGHT CURRENT –GENESIS AT LIVING THINGS.

THESE NEWLY PRODUCED PARTICLE- CLOUD –COMPOUNDS, CONSTRUCT DIFFERENT CNS ELECTRONS, NUCLEONS, PARTICLE COMPOUND CONSTRUCTIONS, UNDER REGENERATIVE – DEGENERATIVE A. S. I. F.P. Mol. C. I. C., MOST OF INCOMING EXOGENOUS S. Y. – F.P. I.I. – p. cl. INTERACT WITH PRE-EXISTING INTERNAL ELECTRON NUCLEON S. Y. – F.P. – I.I. – p. cl. _ COMPOUNDS IN DIFFERENT CNS CENTERS ELECTRONS NUCLEONS.

DURING MEMORIZATION AND DECISION MAKING PHENOMENONS, THE DIFFERENT A. S. I. F.P. Mol. C.I.C. TAKE PLACE EXPONENTIALLY BETWEEN INTERACTING EXOGENOULS INCIDENT PARTICLE CLOUD, WHEN THOSE INTERACTING AND COMPARING, THESE NEWLY INTERED PARTICLE CLOUDS, WITH THOSE PRE-EXISTING PARTICLE CLOUD COMPOUNDS, IN ORDER THROUGH THESE DIFFERENT COMPARISIONS, THE CNS ELECTRONS NUCLEONS FINALLY REACH INTO CONCLUSIONS, AND DECISION MAKINGS, WITH ENTIRE INDIGENOUS PREVIOUSLY CONSTRUCTED EXISTING INFORMATIONS IMAGES PARTICLE CLOUD COMPOUNDS, INSIDE ENTIRE ELECTRONS-NUCLEONS OF DIFFERENT CNS CENTERS.

FINDING BEST SOLUTIONS AND CHOICES THROUGH THESE INNORMOUS DIFFERENT A. S. I. F. P. Mol. C.I.C. THROUGH COMPARING ALL OF PREVIOUSLY RECORDED S. Y. –F.P. – I.I.- p.cl. COMPOUNDS IN DIFFERENT CNS – CENETRS ELECTRONS-NUCLEONS, FINALLY THE HIGHER DECISION MAKING BRAIN CENTERS ELECTRONS – NUCLEONS SELECT BEST SOLUTIONS THROUGH ABOVE RESEARCHES, AND EXECUTE THE PROPER DECISIONS AND RESPONSES FOR ONGOING SITUATION EVENTS, DECIDES, SELECTS, AND EXCUTE PROPER ORDERS ON BASIS OF FINAL FINDINGS, BETWEEN THE DIFFERENT RESEARCHES, BETWEEN THE EXISTING FACTS IN RECORDED INFORMATIONS-IMAGE FORMS, WHICH EXIST IN FORMS OF PARTICLE-COMPOUNDS FORMS FOR COMPARISON OF ENTIRE BRAIN CENTERS, KNOWLEDGES INSIDE ELECTRONS-NUCLEONS, THESE ARE PEOCEDURE USED DURING DECISION MAKING PHENOMENONS, AND MEMORIZING PHENOMENONS, AND EXECUTION AND ORDER ISSUING PHENOMENONS INSIDE DIFFERENT CENTER CNS ELECTRONS –NUCLEONS.

PARTICLE CLOUD TRANSMISSION ROUTES

AIRBORN PARTICLE CLOUD CIRCULATION SYSTEMS CIRCUITS BETWEEN DIFFERENT INDIVIDUALS

==

INTERNAL BODY PARTICLE CLOUD CIRCULATION SYSTEMS

TRANSMISSION ROUTES OF S. Y. E. T. – F.P. – I. I. – P. cl.:

PARTICLES AND PARTICLE CLOUDS POSSESSES TWO MAIN PARTICLE CIRCULATION SYSTEMS: 1- INTERNAL BODY PARTICLE CIRCULATION SYSTEMS, 2 – EXTERNAL BODY PARTICLE CLOUD CIRCULATION SYSTEMS.

1 – EXTERNAL BODY PARTICLE CLOUD CIRCULATION SYSTEMS:

THE PARTICLE CLOUDS MOSTLY TRANSMIT AIRBORN FROM ONE INDIVIDUAL TO OTHER IN CLOSED PARTICLE CLOUD CIRCULATION SYSTEMS (PCS). IN SONIC, ELECTRIC AND LIGHT PARTICLE CLOUDS, WHEN TRANSMISSION BETWEEN

MULTIPLE INDIVIDUAL TAKE PLACE, ALMOST ALL S. Y. –F.P. – I.I. – P. cl. TRANSIT AIRBORN IN CLOSED PARTICLE CLOUD CIRCULATION PATHS AT MOST.

THE INTER - INDIVIDUAL LIVING THINGS PARTICLE TRANSMISSIONS ARE AIRBORN. SOME TIMES THE PARTICLE CLOUDS TAKE DIFFERENT TRANSMISSION ROUTES, MOSTLY INTER INTO BODY THROUGH SENSARY FUNDAMENTAL PARTICLE TRANSIT ROUTES. ALSO THERE IS TRANSMISSION THROUGH FOOD PRODUCTS AS WELL, EVEN THERMAL PARTICLES, ELECTRIC PARTICLES QUICKLY TRANSIT THROUGH SKIN AS WELL. IN PLANTS MOST PARTICLES TRANSIT AIRBORN, TRANSIT THROUGH ROOTS ALSO HAS POSSIBLITIES.

IN EARTH TRANSMISSION OF BIOFRIENDLY SONIC LIGHT FUNDAMENTAL PARTICLE INFORMATION IMAGE PARTICLE CLOUDS (S. Y. – F.P. –I.I. – P. cl.), BETWEEN HUMANS AND ANIMALS MOSTLY TRANSIT AIRBORN BETWEEN DIFFERENT INDIVIDUALS, THE PROCESS IS PARTICLE CLOUD TRANSMISSION SYSTEMS IN CLOSED AIRBORN CIRCULATIONS BETWEEN DIFFERENT INDIVIDUALS, AND PARTICLE CLOUDS AIRBORN TRAVEL BACK AND FORTH BETWEEN HUMAN SPECIES, BOTH NORMAL PARTICLE CLOUDS AS WELL AS ABNORMAL PARTICLE CLOUDS ALL TRANSMIT AIRBORN FROM ONE HUMAN TO OTHER IN AIRBORN PARTICLE CLOUD CIRCULATION SYSTEM CIRCUITS.

2- INTERNAL BODY PARTICLE AND PARTICLE CLOUD CIRCULATION SYSTEMS:

THE MAIN PARTICLE CIRCULATION SYSTEMS INSIDE ANIMAL BODY, CIRCULATE INSIDE INTERNAL NEURAL NANO-DUCT'S PARTICLE CLOUDS CIRCULATIONS SYSTEMS. NEURAL SYSTEMS EITHER IN NANO- LEVELS, OR IN MICRO LEVELS, AND MACRO LEVELS, ALL CIRCULATE INSIDE VOLUNTARY AND NON- VOLUNTARY, CENTRAL AND PERIPHERAL NERVOUS SYSTEMS, PARTICLE CIRCULATION SYSTEMS (PCS).

THE TRANSIT OF S. Y. – F.P. - I.I. - p. cl. INSIDE ANIMALS BODY MOSTLY TRANSIT THROUGH BI-DIRECTIONAL FINE NEURAL FIBERS EFFERENT AND AFFERENT PARTICLE CURRENTS, WHICH CONNECTING DIFFERENT ALL HUMAN ORGANS TOTAL BODY ELECTRON-NUCLEONS PARTICLE POPULATIONS TO CNS TOTAL POPULATION ELECTRON-NUCLEONS IN DIFFERENT CNS CENTERS, THIS IS BIDIRECTIONAL FUNDAMENTAL PARTICLE CURRENTS POPULATIONS CONNECT TO CNS ATOMS TOTAL POPULATIONS IN DIFFERENT CENTERS, THROUGH INTELLECTUAL PARTICLE CENTER SYSTEMS THE DIFFERENT CENTRAL AND PERIPHERAL ELECTRON NUCLEON POPULATIONS COORDINATE PHYSICAL CHEMICAL BIOLOGICAL FUNCTIONS WITH EACH OTHER, THROUGH PARTICLE PRECISION ACCURACIES.

CLASSIFICATION OF PARTICLE CLOUDS

THE S. Y. T. E. – F.P. – I.I. – P. cl. DIVIDE INTO TWO MAJOR CLASSES: 1- NORMAL PARTICLE CLOUDS. 2 – ABNOMRMAL PARTICLE CLOUDS.

THE PSYCHIATRIC DISEASES ARE PRODUCED THROUGH THE TRANSMISSION OF ABNORMAL S. Y. – F.P. – I. I. – P. cl. FROM ONE PARTICLE CLOUD INFLICTED INDIVIDUAL, TO RECIPIENT OTHER INDIVIDUALS. AND PSYCHIATRIC DISEASES ARE PARTICLE CLOUD (S. Y. – F.P. – I.I. –p. cl.) TRANSMITTED DISEASES.

(PSYCHIATRIC DISORDERS ARE ABNORMAL- PARTICLE CLOUD TRANSMITTED DISEASES).

NEWBORN AT BIRTH, DO NOT POSSESSES ANY SIGNIFICANT AMOUNT S. L. T. E. – F.P. – I.I. – P. cl. – COMPOUNDS INSIDE CNS ELECTRONS AND NUCLEONS. THE PARTICLE CLOUDS TRANSMISSION FROM MOTHER'S P.C.S. TO FETUSES BRAIN ELECTRONS AND NUCLEONS, THROUGH FETAL P. C. S. CONNECTIONS TO MOTHER IS NEGLIGIBLE AMOUNT WHEN THE BABY BORN. THE FETAL INTERNAL ELECTRONS NUCLEONS FUNDAMENTAL PARTICLES INFORMATION IMAGE CLOUD CONSTRUCTIONS INSIDE UTERUS IS NEGLIGIBLE, THE S. Y. E. T. – F.P. – I.I. – P.cl. INTERACTIONS, THE STORAGE OR RETRIEVAL PHENOMENONS MOSTLY START AFTER THE BIRTH, AND CONTINUE DURING ALL LIFE LONG.

DURING THE LIFE, ALL THE NORMAL OR ABNORMAL F.P. – I.I.- P.cl. TRANSMIT INTO INDIVIDUALS CNS ELECTRONS NUCLEONS AND CONSTRUCT THE BRAINS INTERNAL ELECTRON - NUCLEON PARTICLE COMPOUND CONSTRUCTIONS. THE NORMAL OR ABNORMAL PARTICLE CLOUD TRANSMISSIONS DURING LIFE AFTER THE BIRTH, PRODUCE AND CONSTRUCT EITHER NORMAL OR ABNORMAL INDIVIDUAL PARTICLE COMPOUND CONSTRUCTIONS AND DEVELOP TO ADVANCED PSYCHE-GENESIS.

ALL OF THE S. Y. E. T. – F.P. – P. cl. OF NORMAL PSYCHE OR ABNORMAL PSYCHE, BEHAVIOR DISORDERS, NEUROSIS, DEPRESSIVE OR ELATED DISORDERS, PSYCHOSIS, NEUROSIS, ANXIETY, COMPULSIVE DISORDERS, NORMAL OR ABNORMAL GENIUSES, DIFFERENT CRIMINAL BEHAVIORS AND CRIMINALS, ETC. ALL THESE DIFFERENT PARTICLE CLOUDS TRANSMIT THROUGH OUT THE LIFE PROGRESSIVELY FROM OTHER INDIVIDUALS AND ENVIRONMENTS INTO NEWBORN INDIVIDUALS CNS, AND CONSTRUCT PARTICLE COMPOUND CONSTRUCTIONS OF BRAINS OF CHILDREN THROUGH USE OF RECEIVED PARTICLE CLOUDS.

THE E. Y. S. T. – F.P. – I.I. – P. cl. TRANSMISSIONS FROM ONE INDIVIDUAL TO OTHER INDIVIDUAL CONSTRUCT THOUGHT CURRENT SYSTEMS WHICH ARE PARTICLE CLOUD CIRCULATION SYSTEMS, AND PSYCHE MOSTLY IS ACQUIRED PHENOMENON, CONGENITAL NEUROLOGICAL DISEASES ARE DIFFERENT ISSUES, ALTHOUGH THOSE ALSO MAY CAUSE MENTAL DISORDERS.

THE ABNORMAL FUNDAMENTAL PARTICLE TRANSMITTED DISEASES, MOSTLY ARE PROTRACTED AND CHRONIC TYPES DISORDERS, FOR BEING INFLICTED BY PARTICLE CLOUDS DISORDERS, MOSTLY REQUIRE PROTRACTED TRANSMISSION TIMES, BUT IN SOME CASES PARTICLE CLOUD TRANSMIT IS FAST AND QUICK, EVEN IN SOME CASES COUPLE EXPOSURES AND PARTICLE CLOUDS TRANSMIT MAY PRODUCE PARTICLE INFLICTIONS.

PSYCHOGENESIS – ABILITIES OF INDIVIDUALS ARE DIFFERENT FROM EACH OTHER, SOME INDIVIDUALS HAVE STRONG ABILITIES TO ALTER THE PSYCHE OF OTHERS. ALSO THERE ARE INDIVIDUALS WHO ARE MORE PSYCHE - SUSCEPTIBLE THAN THE OTHERS. SOME INDIVIDUALS ARE PRONE FOR PARTICLE INFLICTIONS, AND THE OTHERS ARE NOT.

A FEW FACTS ABOUT PARTICLE CLOUD TRANSMISSION:

1- TRANSMISSION OF NORMAL S. Y. - F.P. – I.I. – P. cl. FROM ONE TO OTHER, PRODUCE NORMAL PSYCHE, OR NORMAL INDIVIDUAL.

2 - TRANSMISSION OF ABNORMAL S.Y. –F.P. –I.I. –P. cl. PRODUCE ABNORMAL PSYCHE.

3 - TALKING IS EXCHANGING PARTICLE CLOUDS (S. Y.- F.P. – I.I. – P. cl.) BETWEEN INDIVIDUALS THROUGH EXOGENOUS PARTICLE CLOUD CIRCULATION SYSTEMS, THE PARTICLE CLOUDS CIRCULATE BETWEEN INDIVIDUALS IN PARTICLE CLOUD CIRCULATION SYSTEMS CIRCUITS.

4- TRANSMISSION OF NORMAL PARTICLE CLOUDS FROM NORMAL PARTICLE CLOUD DONERS TO NORMAL RECIPIENTS. PRODUCE NORMAL PSYCHE.

5 - TRANSMISSION OF ABNORMAL PARTICLE CLOUDS FROM SICK DONERS TO NORMAL RECIPIENTS, PRODUCE SICK PSYCHE MOSTLY.

6 - ABNORMAL S. Y. T. E. – F.P. – I.I. – P. cl. TRANSMISSION IS CAUSE OF MENTAL DISORDERS

7 - PSYCHIARTRIC DISEASES ARE ABNORMAL PARTICLE CLOUD TRANSMITTED DISORDERS.

8- NORMAL PARTICLE CLOUD COMBINATION WITH INTERNAL ELECTRONS NUCLEONS PARTICLE COMPOUNDS PRODUCE NORMAL CONSTRUCTION CNS ELECTRONS NUCLEONS. AND NORMAL S.Y. – F.P. – I.I. – P.cl. WILL CIRCULATE INSIDE CNS ELECTRONS NUCLEONS PARTICLE CIRCULATION SYSTEMS CENTER.

9- ABNORMAL PARTICLE CLOUDS COMBINATIONS WITH CNS ELECTRON NUCLEONS PARTICLE COMPOUNDS CONSTRUCT ABNORMAL CNS PARTICLE CLOUD COMPOUND CONSTRUCTIONS ELECTRON NUCLEON. MANIFEST ITSELF AS ABNORMAL SYMPTOMS AND SIGNS.

CAUSE OF PSYCHIATRIC DISOREDERS ARE PROLONG TRANSMISSION OF ABNORMAL S. Y. – F.P. – I.I. – P. cl. FROM ABNORMAL PARTICLE CLOUD INFLICTED SICK PSYCHIATRIC PATIENTS, TO NORMAL CHILDREN, STUDENTS, INDIVIDUALS, THE PARTICLE-CLOUD DONER AND TRANSMITTER INDIVIDUALS CAN BE ANY ONE SUCH AS: PARENTS, TEACHERS, INSTRUCTERS, RULERS, SOCIO – POLITICAL OR RELIGIOUS INDIVIDUALS, ETC.

THE ABNORMAL S. Y. – F.P. – I.I. – P. cl. AIRBORN TRANSMIT FROM SICK TO NORMAL INDIVIDUALS RECIPIENT'S SENSARY ORGANS, ELECTRONS, NUCLEONS ATTRACT PARTICLE CLOUDS, AND SENSARY ORGANS THROUGH NANO-FIBERS OF NEURAL CONSTRUCTIONS CARRY INCOMING ABNORMAL PARTICLE CLOUDS INTO RECIPIENTS CNS SENSARY CENTERS ELECTRONS NUCLEONS, AND CONSTRUCT THE PARTICLE CLOUD COMPOUND CONSTRICTIONS IN DIFFERENT CNS CENTERS ELECTRONS NUCLEONS CONSTRUCTIONS.

INCOMING PARTICLE CLOUDS INTER INTO CNS SENSARY INTERNAL ELECTRONS, NUCLEON SUBSYSTEM-UNITS CHEMICAL LAB.S, AND THE EXOGENOUS PARTICLE CLOUDS COMBINE WITH INTER ELECTRON NUCLEON PARTICLE COMPOUNDS, AND PRODUCE S. Y. – F.P. – I.I. – P. cl. COMPOUNDS AND CONSTRUCT RECIPIENTS ELECTRONS NUCLEONS PARTICLE COMPOUNDS CONSTRUCTIONS, WITH INCOMING ABONORMAL PARTICLE CLOUD COMPOUNDS, ABNORMAL SONIC PARTICLE CLOUDS CONSTRUCT ABNORMAL SONIC PARTICLE COMPOUNDS (S. F.P. – I.I. – P. cl. _ COMPOUNDS) IN AUDITARY CNS CENTERS ELECTRONS NUCLEONS ABNORMALLY, AND ABNORMAL LIGHT PARTICLE CLOUDS CONSTRUCT ABNORMAL LIGHT PARTICLE COMPOUNDS IN VISION CENTERS ELECTRONS NUCLEONS AND CONSTRUCT THE VISION CENTERS INTER ELECTRON-NUCLEON PARTICLE COMPOUND CONSTRUCTIONS WITH ABNORMAL Y. F.P. – I.I. – P. cl. COMPOUNDS,

THE INTERACTIONS OF THESE ABNORMAL S. Y. – F.P. – I.I. – P. cl. WITH EACH OTHER, BETWEEN DIFFERENT CNS CENTERS, AS WELL AS THEIR INTERACTIONS WITH INCOMING OTHER S. Y. – F.P. – I.I. – P. cl. PRODUCE ABNORMAL PARTICLE CLOUD CURRENTS, THE ABNORMAL PARTICLE CLOUD INTERACTIONS TAKE PLACE THROUGH A. S. I. F.P. – Mol. C.I.C. AND PRODUCE ABNORMAL THOUGHT CURRENTS, AND ABNORMAL PSYCHE, THIS IS PHENOMENON OF ABNORMAL PSYCHE- GENESIS AND ABNORMAL THOUGHT CURRENT –GENESIS PHENOMENONS,

PSYCHIATRIC (PARTICLE CLOUD TRANSMITTED) DISEASES TREATMENT

PARTICLE TRANSMITTED DISEASES AND THEIR TREATMENTS ARE BIG SUBJECTS, HERE ESSENTIALS OF TREATMENT IN COUPLE LINES ARE AS FOLLOWING:

1 - THE STORED ABNORMAL S.Y.E.T.- FP- I.I. – P. Cl., MUST BE REMOVED AND DISCARDED TO OUT OF THE CNS INTER – ELECTRON, NUCLEON PARTICLE CLOUD- COMPOUND CONSTRUCTIONS, FROM TOTAL CNS ELECTRONS, NUCLEONS POPULATIONS FROM ALL DIFFERENT CNS CENTERS. THIS PROCESS ACHIEVED THROUGH THE DEGENERATIVE A. S. I. FP. I. Mol. C.I.C. WHICH THIS PROCESS IS EQUAL TO RETRIEVAL, REMOVAL AND ELIMINATIONS OF THE ALL ABNORMAL CLOUDS, TO OUT OF CNS ELECTRONS, NUCLEONS PARTICLE COMPOUND CONSTRUCTIONS.

2 - THEREAFTER THROUGH REGENERATIVE A. S. I. FP. Mol. C.I.C. IN THEIR PLACE, THE ENTIRE CNS ELECTRONS, NUCLEONS TOTAL POPULATIONS ALL MUST BE RECONSTRUCTED THROUGH THE USE OF NORMAL EXOGENOUS S.Y.E.T.-FP.- I.I. –P. Cl., AND THESE NORMAL CLOUDS ALL STORED INSIDE THE CNS- ELECTRONS AND NUCLEONS, AT THE NORMAL PARTICLE CLOUD COMPOUND CONSTRUCTION FORM, (PROCESS OF THE NORMAL CNS PARTICLE CLOUD COMPOUND ELECTRONS NUCLEON RECONSTRUCTIONS). DETAIL PROCEDURES DISCUSSED INOTHER TEXTS.

PSYCHE-GENESIS

DIFFERENT KIND PARTICLE CLOUD (S. Y. –F.P. – I.I. – P. cl.) TRANSMISSIONS, PRODUCE DIFFERENT KIND PSYCHE, AND DIFFERENT KIND THOUGHT CURRENTS

RMAL PARTICLE CLOUD DONNERS, CONSTRUCT CNS INTERNAL ELECTRON-NUCLEON PARTICLE COMPOUND CONSTRUCTIONS OF PARTICLE CLOUD RECIPIENT, CHILDREN, STUDENTS, INDIVIDUALS INTERNAL ELECTRON-NUCLEON PARTICLE COMPOUND CONSTRUCTIONS, WITH ABNORMAL S. Y. – F.P. –I.I. – P. cl. – COMPOUND CONSTRUCTIONS.

THE PSYCHOTIC DELUSIONAL PARTICLE CLOUD (S.Y. – F.P.- I.I. – P. cl.) TRANSMISSION, TO RECIPIENTS PRODUCE PSYCHOSIS TYPES PARTICLE CLOUD COMPOUND CONSTRUCTIONS, INSIDE RECIPIENTS CNS ELECTRONS NUCLEONS. THE DEPRESSIVE PARTICLE CLOUD (S.Y. – F.P. – I.I. – P. cl.) TRANSMISSIONS PRODUCE DEPRESSION TYPE PARTICLE CLOUD INFLICTIONS ATOMS. THE NEUROTICS NERVOUS PARTICLE CLOUD TRANSMITTERS TRANSMIT NEUROSIS TYPES PARTICLE CLOUDS TO RECIPIENTS AND CONSTRUCT NEUROTIC KIND PARTICLE CLOUD COMPOUNDS INSIDE THE RECIPIENTS CNS INTERNAL ELECTRONS NUCLEONS PARTICLE COMPOUNDS. AND SO ON,

THE NORMAL PSYCHE PARENTS, TEACHERS, INSTRUCTORS, TEACH AND TRANSMIT NORMAL PARTICLE CLOUDS (S. Y. – F.P. – I.I. – P. cl.) TO RECIPIENT CHILDREN, STUDENTS, INDIVIDUALS, AT HOME, SCHOOL, UNIVERSITY, AND PUBLIC PLACES, AND PRODUCE NORMAL CNS S. Y. – F.P. –I.I. – P. cl. -PARTICLE COMPOUND COMSTRUCTIONS IN RECIPIENT CHILDRENS, STUDENTS, AND INDIVIDUALS,

THE DEVIATED AND CRIMINAL PARENTS, TEACHERS, SCHOOLS, POLITITIONS, UNIVERSITIES, INSTRUCTORS, TRANSMIT DEVIATED AND CRIMINAL TYPES OF PARTICLE CLOUDS (S. Y. – F.P. – I.I. – P. cl.) TO RECIPIENT STUDENTS, CHILDREN, PARTIES, SOCIO-POLITICAL SOCIETIES, AND INFLICT NORMAL INDIVIDUALS WITH ABNORMAL DEVIATED OR CRIMINAL PARTICLE CLOUDS (S. Y. – F.P. – I.I. – P. cl.) AND CONSTRUCT RECIPIENT NORMAL OR ABNORMAL INDIVIDUALS INTER CNS INTER ELECTRON, NUCLEON, PARTICLE COMPOUND CONSTRUCTIONS, WITH ABNORMAL DEVIATED TYPES S.Y. – F.P. – I.I. – P. cl. –COMPOUND CONSTRUCTIONS AND PRODUCE CRIMINALS AND DEVIATES,

SOME INDIVIDUALS POSSESSES REMARKABLE PSYCHO-GENESIS POWERS AND CAPABILITIES THAN OTHER, SOME ABNORMAL PARTICLE CLOUD INFLICTED INDIVIDUALS, WILL TRANSMIT MENTAL DISEASE TO OTHERS INDIVIDUALS, EVEN IN SHORT CONTACT, IN CONTRAST IN MOST OTHERS TRANSMISSION IS VERY SLOW PROCESS, PSYCHE-GENESIS CAPABILITIES OF DIFFERENT INDIVIDUALS ARE DIFFERENT FROM EACH OTHER.

MOST INDIVIDUALS TRANSMIT THE KIND OF S.Y – F.P. – I.I. –P. cl. FROM SELF, WHICH THEIR INTERNAL CNS INTER ELECTRONS –NUCLEONS PARTICLE CLOUD COMPOUNDS CONSTRUCTIONS, HAVE BEEN MADE AND CONSTRUCTED FROM THOSE KIND S.Y. –F.P. – I.I. – P. cl. COMPOUNDS.

THE INDIVIDUALS WHOSE INTERNAL CNS INTER ELECTRON-NUCLEON PARTICLE COMPOUND CONSTRUCTIONS, HAS BEEN PRODUCED BY USE OF PSYCHOTIC S. Y. – F.P. – I.I. – P. cl. COMPOUNDS CONSTRUCTIONS, THESE PATIENTS ARE PSYCHOSIS TRANSMITTERS, THE PSYCHOSIS TRANSMITTERS MOSTLY PRODUCE AND TRANSMIT, AIRBORN ABNORMAL PSYCOSIS PARTICLE CLOUDS (S.Y. – F.P. – I.I. P. cl.) AND CAUSE PRODUCTIONS OF PSYCHOSIS IN RECIPIENTS, CHILDRENS, AND INDIVIDUALS.

IN CONTRAST THE PARENTS, TEACHERS, INSTRUCTORS, WHOSE CNS INTER ELECTRON – NUCLEON PARTICLE COMPOUND CONSTRUCTIONS, THAT HAVE BEEN CONSTRUCTED FROM DEPRESSIVE, OR ANXIETY NERVOUS TYPES, OR OTHER KIND PARTICLE CLOUDS (S.Y. – F.P. – I.I. P. cl.) COMPOUNDS CONSTRUCTIONS, THESE INDIVIDUALS TRANSMIT FROM SELF DEPRESSION, NEUROSIS, AND ANXIETY KIND PARTICLE CLOUDS TO OTHERS, AND IN RECIPIENTS UNDER A. S. I. F.P. Mol. C.I.C. INSIDE THEIR CNS ELECTRONS NUCLEONS PARTICLE CLOUDS COMBINE WITH INTERNAL ELECTRON NUCLEON PARTICLE COMPOUNDS, AND PRODUCE DEPRESSIVE NEW (S.Y. – F.P. – I.I. – P cl.) PARTICLE CLOUD – COMPOUNDS, OR NEUROTIC S.Y. –F.P. – I.I.—P.cl. COMPOUNDS AT INTER ELECTRON-NUCLEON CONSTRUCTIONS, AND CREATE IN RECIPIENTS THE SAME KIND PARTICLE CLOUD TRANSMITTED DISEASES AND THESE ARE NEW PARTICLE CLOUD INFECTED PATIENTS.

PARTICLES, PARTICLE- CLOUDS, DIFFERENT TYPES NORMAL OR ABNORMAL TYPES FUNDAMENTAL PARTICLE INFORMATION AND IMAGE PARTICLE CLOUDS (F.P. – I.I.- P. cl.) TRAVEL AND TRANSIT AIRBORN, INFLICT RECIPIENT LIVING THINGS, WITH SAME KIND NORMAL OR ABNORMAL PARTICLE CLOUD TRANSMITTED (S. Y. – F.P. – I.I. – P. cl. TRANSMITTED INFECTIONS, PRESENTLY KNOWN AS MENTAL CONDITIONS.

THE MENTAL DISORDERS ARE PARTICLE TRANSMITTED DISEASES, AND PARTICLE INFECTED PATIENTS.

ANIMAL GENESIS

CONSTRUCTIONS OF BIOLOGICALLY NEEDED BODY SYSTEMS AND ORGANS FOR LIVING THINGS, UNDER THE MOLECULAR EVOLUTION ORDERS, AND THROUGH A.S.I. Mol. C.I.C.

DURING THE PARTICLE MOLECULAR EVOLUTION, THE PARTICLE INTELLIGENCE SYSTEMS INVENTED AND CONSTRUCTED DIFFERENT BIOLOGICAL DEVICES, CONSTRUCTED DIFFERENT BODY ORGANS, AND BODY SYSTEMS, PARTICLES FOR PERFORMING ENORMOUS DIFFERENT KINDS EXISTING LIVING THINGS DIFFERENT BIOLOGICAL FUNCTIONS, WHICH THESE TASKS ARE NECESSARY AND NEEDED FOR OPERATION OF A BIOLOGICAL MACHINES.

THE PARTICLES CONSTRUCTED NEEDED BODY SYSTEMS SUCH AS: FLUIDS -FOODS PUMPERS CIRCULATIONS SYSTEMS BIO-DEVICES, BONY – MUSCULAR MOVEMENT BIO- DEVICES, DEFLATING –INFLATING GAS EXCHANGE BIO DEVICES, AND MANY OTHER NEEDED BODY SYSTEMS AND ORGANS WHICH EACH ACHIEVE ONE OR MANY DIFFERENT TASKS, ALL OF BODY SYSTEMS AND ORGANS CONSTRUCTED THROUGH PARTICLE INTELLIGENCE SYSTEMS, AND ATOMS, BIOMOLECULES, CELL'S INTELLIGENCE SYSTEMS COOPERATING WITH EACH OTHER, UNDER THE MOLECULAR EVOLUTION ORDERS, AND A.S. I. F.P. Mol. C.I.C.

FOR COMMUNICATIONS OF DIFFERENT INDIVIDUALS CNS ELECTRONS NUCLEONS WITH EACH OTHER, THE PARTICLES CONSTRUCTED BIO-DEVICES: SUCH AS AIRBORN EXOGENOUS PARTICLE CLOUD CIRCULATION SYSTEMS (EX. -PCS), WHICH EXOGENOUS PARTICLE CIRCULATION SYSTEMS ENABLE DIFFERENT LIVING THING SPECIES

EXCHANGE KNOWLEDGES BETWEEN EACH OTHER, ALSO PARTICLES CONSTRUCTED INDIGENOUS PARTICLE CIRCULATION SYSTEMS (PRESENTLY KNOWN AS NERVOUS SYSTEMS). THE ELECTRONS NUCLEONS OF DIFFERENT LIVING THINGS SPECIES THROUGH THESE PCS – BIO-DEVICES COMMUNICATE WITH EACH OTHER.

NANO- COMMUNICATION BETWEEN TOTAL BODY ELECTRONS – NUCLEONS POPULATIONS

PARTICLE CLOUS (S. Y. E. T. – F.P. – I.I. – P. cl.) CIRCULATION SYSTEMS

THERE ARE TWO CLASSES PARTICLE CLOUD CIRCULATING SYSTEMS:

1- INTERNAL BODY PARTICLE CLOUD CIRCULATION SYSTEMS OR INDIGENOUS PARTICLE CLOUD CIRCULATION SYSTEMS: THIS SYSTEM PARTICLE CLOUD CIRCULATION SYSTEM FLOWS, AND INTER-CONNECTS THE TOTAL BODY ELECTRON-NUCLEON POPULATIONS TO EACH OTHER, IT IS NANO-CIRCULATION SYSTEMS, IT'S MICRO AND MACRO CONSTRUCTIONS FORMS ARE PRESENTLY KNOWN AS NERVOUS SYSTEM TODAY.

2 – THE EXOGENOUS PARTICLE CLOUD CIRCULATION SYSTEMS: THIS IS THE PARTICLE CLOUD (S. Y. E. T. –F.P. – I.I. – P. cl.) CURRENTS CIRCULATION SYSTEMS THAT FLOW AIRBORN BACK AND FORTH BETWEEN CNS ELECTRON NUCLEONS OF DIFFERENT COMMUNICATING INDIVIDUALS, WHICH BY ORDINARY PEOPLE ARE KNOWN AS TALKING AND LISTENING PROCEDURES. THIS EXOGENOUS PARTICLE CLOUD CIRCULATION SYSTEM ALSO EXIST BETWEEN THE OTHER LIVING THINGS DIFFERENT SPECIES AS WELL. THERE ARE THIRD OTHER TYPES PARTICLE CIRCULATION SYSTEMS BETWEEN NON-LIVE NANO-UNITS ALSO.

EXOGENOUS PARTICLE CLOUD CIRCULATION SYSTEMS

AND PHENOMENON OF PARTICLE CLOUD EXCHANGE SYSTEMS BETWEEN DIFFERENT CREATUTES

ANIMALS CNS ELECTRON-NUCLEON THROUGH BILLIONS TO TRILLIONS CONCURRENT DIFFERENT A. S. I. F.P. Mol. C.I.C. WHICH TAKES PLACE BETWEEN DIFFERENT CNS CENTERS ELECTRONS NUCLEONS WITH EXPONENTIAL SPEEDS, PRODUCE AND EMITS AIRBORN PARTICLE CLOUDS TO THE OUTSIDE WORLD INTO AIR, THESE PARTICLE CLOUDS TRANSIT IN ALL AIR DIRECTIONS INTO SPACE.

THE RECIPIENTS CNS ELECTRONS NUCLEONS CAPTURE THESE AIRBORN PARTICLE CLOUDS FROM AIR THROUGH EXISTING MULTI DIFFERENT ORGANS AND SYSTEMS, THESE INCOMING S.Y. – F.P. – I.I. – P.cl. INSIDE RECIPIENTS CNS ELECTRONS NUCLEONS ANALIZED, THROUGH TRILLIONS OF DIFFERENT A. S. I. F.P. Mol. C.I.C. WHICH TAKE PLACE WITH EXPONENTIAL SPEEDS OF INTERACTIONS, BETWEEN DIFFERENT CNS CENTERS ELECTRONS AND NUCLEONS, PRODUCE RESPONSE PARTICLE – CLOUDS AND EMIT THOSE RESPONSE - PARTICLE - CLOUDS (S.Y – F.P. – I.I. – P. cl.) INTO OUTSIDE WORLD AS AIRBORN PARTICLE CLOUDS AGAIN. THESE PARICLE CLOUDS CAPTURED BY OTHERS CNS ELECTRONS-NUCLEONS, THESE BACK AND FORTH PARTICLE CLOUD CURRENTS IN CLOSED PARTICLE CLOUD CIRCULATION SYSTEMS CIRCUITS. CIRCULATE BETWEEN DIFFERENT INDIVIDUALS, ANIMALS AS WELL AS NON LIVE ELECTRONS NUCLEONS. WHICH THESE PHENOMENONS ARE EXOGENOUS PARTICLE CLOUD CIRCULATIONS, AND PARTICLE CLOUD EXCHANGE SYSTEMS, AND PRESENTLY PEOPLE CALL IT SPEAKING AND LISTENING.

ABOVE TYPES FUNDAMENTAL PARTICLE CLOUD CIRCULATING SYSTEM ARE EXTERNAL AIRBORN S. Y. – F.P.- I.I. – P. cl. EXCHANGES, BETWEEN TWO OR MORE DONER AND RECIPIENT LIVING THINGS, SUCH AS HUMAN, OR OTHER ANIMAL SPECIES, ALSO THIS PHENOMENON OCCURS BETWEEN NON- LIVE ELECTRONS NUCLEONS PARTICLE TRANSMISSIONS THROUGH AIRBORN PARTICLE CIRCULATIONS SYSTEMS,

3 - THE THIRD SEGMENTS OF EXOGENOUS PARTICLE CIRCULATION SYSTEMS, AND INDIGENOUS PARTICLE CLOUD CIRCULATING SYSTEMS, ARE CONNECTIONS BETWEEN THESE TWO SYSTEMS, WHICH TAKE PLACE BETWEEN TWO ABOVE EXPLAINED PARTICLE CLOUD CIRCULATION SYSTEMS IN JUCTURES, WHEN BOTH SYSTEMS PARTICLE CLOUD CURRENTS ARE CONNECTING BY SYNAPSUS AND JOINING TO EACH OTHER, AND PRODUCING A SINGLE NON INTERRUPTED PARTICLE CLOUD CURRENT, AS A SINGLE CONNECTED CLOSED CIRCUITS.

PSYCHE GENESIS AND THOUGHT CURRENT GENESIS PHENOMENON

STORING VIRTUAL PARTICLE CLOUD COPIES OF OUTSIDE WORLD INSIDE CNS ATOMS:

(S. Y. –F.P.- I.I.- P.cl. ARE VIRTUAL PARTICLE CLOUD COPIES OF OUTSIDE WORLD'S -THINGS)

THE CNS INTERNAL ELECTRON NUCLEON INTERACTIONS ABOUT STORED S. Y.-F.P.- I.I.- P.cl. OR VIRTUAL CLOUDS WHICH ALL ARE INFORMATIONS AND IMAGES ABOUT OUTSIDE WORLD SUBJECTS, THESE INTERACTIONS BETWEEN CNS ELECTRONS NUCLEONS MIMICS MOVIES LIKE FUNCTIONS, FROM ACTING, SPEAKING, MOVING, TEACHING, FROM OUTSIDE SUBJECTS, BIOLOGICALLY SENSE AS PSYCHE, THOUGHT CURRENTS.

SONIC AND LIGHT FUNDAMENTAL PARTICLES CONSTRUCT INFORMATION IMAGE PARTICLE CLOUDS COPIES FROM EXISTING THINGS OF EARTH (S.Y. – F.P. – I.I. –P. cl.), SONIC PARTICLES AND LIGHT PARTICLES BOTH ARE SENSIBLE, AND WE CAN SEE LIGHT PARTICLES AND WE CAN HEAR SONIC PARTICLES OF EARTH. THE PRODUCED SONIC PARTICLE AND LIGHT PARTICLE INFORMATION IMAGE PARTICLE CLOUDS (S. Y. – F.P. – I.I.- P.cl.), WHEN THEY CONSTRUCT THE S. Y. – F.P. – I.I. – P. cl. COMPOUNDS INSIDE CNS ELECTRONS AND NUCLEONS, WHEN THEIR S.Y. – F.P. – I.I. – P.cl CURRENTS INTERACTIONS UNDER A. S. I. F.P. Mol. C.I.C., INSIDE THE BRAIN ELECTRONS NUCLEONS, BIOLOGICALLY SENSED AND PRODUCE THOUGHT CURRENT SYSTEMS AND PSYCHE, THIS IS PHENOMENON OF PSYCHE-GENESIS AND THOUGHT CURRENT –GENESIS.

SONIC – LIGHT FUNDAMENTAL PARTICLE INFORMATION IMAGE PARTICLE –CLOUDS (S.Y.-F.P.- I.I.-P.cl.) CONSTRUCT S. Y. – F.P. – I.I. – PARTICLE CLOUD COPIES FROM EXISTING THINGS OF EARTH, THE EARTH'S LIGHT PARTICLES AND SONIC PARTICLES ARE SENSIBLE.

THE S. Y. – F.P. – I.I. – P. cl. –COMPOUNDS WHICH ARE COPIES OF ENVIRONMENTAL SUBJECTS AND EVENTS INSIDE CNS INTER-ELECTRON NUCLEON IN PARTICLE COMPOUND CONSTRUCTIONS FORMS, AND CONSTRUCT CNS ELECTRON NUCLEONS, THEIR INTERACTIONS IN BRAIN CAN BE SENSED AS EXISTING ORIGINAL SUBJECTS AND EVENTS PLAYING THROUGH THOUGHT CURRENTS AND PSYCHE FORMS. THESE INFORMATION IMAGE PARTICLE CLOUD COPIES INSIDE CNS ELECTRONS NUCLEONS AND THEIR INTERACTIONS WITH EACH OTHER UNDER A. S. I. F.P. Mol. C.I.C., BIOLOGICALLY SENSE AS THOUGHT CURRENTS AND PSYCHE.

COMBINING S.Y. – F.P. – I.I. –P. cl. WITH SILICON'S INTERNAL ELECTRON-NUCLEON PARTICLE COMPOUNDS

STORAGE OF PARTICLE CLOUDS INSIDE SILICON ATOMS.

THIS IS INFORMATION IMAGE PARTICLE CLOUD - STORAGE PHENOMENON, INSIDE SILICON'S NON LIVE ELECTRONS NUCLEONS, IN THE FORMS OF PARTICLE CLOUD COMPOUNDS.

THROUGH COMPUTER TECH. TRANSMIT THE PRODUCED SONIC LIGHT FUNDAMENTAL PARTICLE INFORMATIONS IMAGES PARTICLE CLOUDS (S. Y. –F.P. – I.I. – P. cl.) WHICH ARE PARTICLE CLOUD COPIES FROM ENVIRONMENTAL EXISTING THINGS, INTO SUBSYSTEM UNITS CHEMICAL LAB.S OF SILICON ELECTRONS AND NUCLEONS, UNDER

REGENERATIVE A. S. I. F.P. Mol. C.I.C. COMBINE EXOGENOUS INCOMING S. Y. –F.P. – I.I. –P.cl. WITH SILICONS INTER-ELECTRON- NUCLEON PARTICLE COMPOUNDS, AND PRODUCE SONIC – LIGHT FUNDAMENTAL PARTICLE INFORMATION IMAGE PARTICLE CLOUD COMPOUNDS (S. Y. –F.P. – I.I. – P.cl. _ COMP.) THIS PHENOMENON IS STORAGE OF EXOGENOUS S. Y.- F.P. – I.I. – P. cl. INSIDE SILICONS INTER ELECTRONS NUCLEONS, IN THE SONIC – LIGHT FUNDAMENTAL PARTICLE INFORMATION IMAGE PARTICLE CLOUD COMPOUND MOLECULAR FORMS.

THE S. Y. – F.P. – I.I. – P. cl. RETRIEVAL FROM SILICON ELECTRON – NUCLEONS

PHENOMENON OF RETRIEVAL OF RECORDED MEMORIES

REVERSE OF ABOVE INTERACTIONS, UNDER DEGENERATIVE A. S. I. F.P. Mol. C.I.C., CAUSE BREAK DOWN OF LARGE S.Y.- F.P. – I.I.- P.cl. _COMPOUNDS, INTO SMALL MOLECULAR CONSTRUCTING MOLECULES, CAUSE RELEASES OF INFORMATIONS IMAGE PARTICLE CLOUDS TO OUT OF SILICON'S ELECTRONS NUCLEONS INTO AIR, AND AFTER RELEASE OR RETRIEVAL, THE S. Y. –F.P. – I.I. – P.cl. CAN BE PRESENTED ON TV SCREENS. AND THIS IS PHENOMENON OF RETRIEVAL AND RELEASE OF PARTICLE CLOUDS INTO OUTSIDE SPACE.

COMMUNICATION OF ATOMS, ELECTRONS, NEUTRONS, PROTONS AND OTHER NANO – UNITS, WITH EACH OTHER

NANO NEUROLOGY OF ELECTRONS, PROTONS, NEUTRONS, ATOMS AND NANO- UNITS.

PARTICLE CIRCULATION SYSTEMS (PCS) INSIDE ELECTRONS, NUCLEONS, ATOMS, AND NANO – UNITS.

1- THE AIRBORN, AND PERIPHERAL ATOM SPACE (PAS) PARTICLE CIRCULATION SYSTEMS, (A.- P. C. S.)

THE AIRBORN PARTICLE CIRCULATION SYSTEMS AND PERIPHERAL ATOM PARTICLE CIRCULATION SYSTEMS (A-PCS & PAS – PCS) ARE PARTICLE CURRENTS, FUNDAMENTAL PARTICLE CIRCULATION SYSTEMS, AND FUNDAMENTAL PARTICLE CLOUD CIRCULATIONS SYSTEMS, AS WELL AS FUNDAMENTAL PARTICLES INFORMATIONS AND IMAGES PARTICLE CLOUD CIRCULATION SYSTEMS, WHICH CIRCULATE MOSTLY IN CLOSED PARTICLE AND PARTICLE CLOUD CIRCULATIONS CYCLES, AND SOME OF THEM ARE SIMPLE FUNDAMENTAL PARITCLE CURRENTS, AND PARTICLE CLOUD OR INFORMATION – IMAGE PARTICLE CLOUD CIRCUITS AND CURRENTS.

BUT THE MOST OF THE A – PCS & PAS – PCS, EITHER CLOSED PARTICLE CIRCUIT CYCLES, OR OPEN PARTICLE CLOUD CURRENTS CONNECTS WITH THE INTERNAL ELECTRON, INTERNAL NUCLEON, INTERNAL ATOMS, OR INTERNAL NANO-UNITS PARTICLE CIRCULATION SYSTEMS (I- P. C. S.) AT CLOSED CYCLES, AND BOTH EXTERNAL NANO – UNITS FP – I.I. – P. Cl. CIRCULATION SYSTEMS, AND INTERNAL NANO – UNIT FP – I.I. – P. Cl. CIRCULATION SYSTEMS CONNECT WITH EACH OTHER, AND THROUGH THESE PARTICLE CLOUD CIRCULATIONS THE DIFFERENT ELECTRONS, NEUTRONS, PROTONS, AND ATOMS AND ANOTHER NANO- UNITS ALL OVER UNIVERSE COMMUNICATE WITH EACH OTHER.

THE P. C. S. INSIDE NANO- UNITS:

1 - PCS INSIDE PERIPHERONS.

2 - PCS INSIDE NUCLEUS.

3 - PCS INSIDE ELECTRONS.

4 - PCS INSIDE THE NEUTRONS.

5 – PCS INSIDE THE PROTONS.

6 – PCS INSID QUARKS (ANOTHER KIND NANO – UNIT).

7 - PCS INSIDE INTERNAL ATOMS SUB - SYSTEMS, SUBSYSTEM – UNITS, AND SYSTEMS, WHICH ALL ARE DIFFERENT KIND NANO- UNIT CONSTRUCTIONS INSIDE ELECTRONS, NEUTRONS, PROTONS, ATOMS.

P.C.S. INSIDE THE MICRO – UNITS, JOINT NANO – UNIT AND MICRO-UNIT P.C.S.

THE CHROMOSOMAL CONSTRUCTION, CELL NUCLEUS, CELL CYTOPLASM, LARGE BIO - MOLECULAR CONSTRUCTIONS, THE INTERNAL CYTOPLASMIC AND INTERNAL NUCLEUS DIFFERENT ORGANS AND SYSTEMS ALL ARE EXAMPLES FROM MICRO –UNITS:

EXAMPLES FROM INTERNAL CYTOPLASMIC ORGANS AND SYSTEMS ARE:

MITOCHONDRIA, CELL MEMBRANE, CYTOPLASMIC TRANSPORTATION SYSTEMS, CYTOPLASMIC PCS, CYTOPLASMIC IMMUNITY SYSTEMS, CYTOPLASMIC CHEMICAL LAB.S, CYTOPLASMIC REPRODUCTION AND DUPLICATION SYSTEMS, CYTOPLASMIC CENTRAL INTELLIGENCE SYSTEM CENTERS, PCS, AND PIS OF THE CYTOPLASM, ETC.,

THE INTERNAL CELL NUCLEI DIFFERENT ORGANS AND SYSTEMS EXAMPLES ARE:

THE CHROMOSOMAL BIOMOLECULAR AND MOLECULAR DUPLICATION AND DIVISION SYSTEMS AND ORGANS, CELL NUCLEI - INTELLIGENCE SYSTEM CENTERS, NUCLEI MOLECULAR TRANSPORTATION ORGANS AND SYSTEMS, NUCLEAR DEFENSIVE AND IMMUNITY CENTERS SYSTEMS AND ORGANS, CELL NUCLEUS PIS, CELL NUCLEUS PCS, ETC.

THE MOST CELLS, AND MANY CELL DIFFERENT ORGANS AND CELL SYSTEMS CAN BE CLASSIFIED UNDER MICRO-UNIT CONSTRUCTIONS, AND ALL ATOM – BASE MOLECULAR AND BIO-MOLECULAR CONSTRUCTIONS IN EARTH ALL ARE PART OF THE MICRO- UNITS DIFFERENT CLASS SYSTEMS.

EVOLUTION IN NANO – UNITS:

THE A. S. I. FP. Mol. C.I.C. BETWEEN THE NANO – UNITS AND DIFFERENT FUNDAMENTAL PARTICLE MOLECULAR STRUCTURES UNDER THE MOLECULAR EVOLUTION ORDERS AND PATHS, PRODUCED LARGR NANO – UNIT STRUCTURES AND CAUSED PRODUCTIONS OF THE LARGER BIOMOLECULAR CONSTRUCTIONS.

EVOLUTION IN MICRO- UNITS:

UNDER AUTONOMOUS INTER ATOM – BASE AND INTER - BIOMOLECULAR BASE AUTONOMOUS SEQUENTIAL CHEMICAL INTERACTION CHAINS AND CYCLES, THROUGH MOLECULAR EVOLUTION ORDERS AND PATHS, THE ATOM AND BIOMOLECULAR BASE MOLECULAR EVOLUTION AND AUTONOMOUS CHEMICAL INTERACTIONS BETWEEN DIFFERENT MOLECULARE STRUCTURES, FINALLY CONSTRUCTED DIFFERENT CELLS, TISSUES, BODY ORGANS, AND BODY SYSTEMS, AND LIVING THINGS.

INTERNAL CELL ORGANS AND INTER- CELL SYSTEM'S PARTICLE INTELLIGENCE SYSTEMS (PIS), AND PARTICLE CIRCULATION SYSTEMS (PCS).

IN ANY GIVEN MICRO-UNIT (FOR EXAMPLE A CELL), THE EXISTING INDEPENDENT ATOMS, ELECTRONS, NEUTRON, PROTON'S, NANO- UNITS INSIDE THE MICRO-UNIT, ALL POSSESSES INDEPENDENT PARTICLE INTELLIGENCE SYSTEM CENTERS (OR NANO-UNIT PIS), AND THE MOLECULAR CONSTRUCTIONS OF THE ENORMOUS DIFFERENT ATOM-BASE BIO-MOLECULAR CONSTRUCTIONS OF DIFFERENT INTERNAL CELL EXISTING DIFFERENT MICRO- UNIT STRUCTURES, INSIDE ANY GIVEN CELL OR MICRO-UNIT'S INTERNAL NUCLEUS OR INTERNAL CYTOPLASMIC ORGANS AND SYSTEMS THESE ALL ALSO POSSESSES ADDITIONAL ANOTHER DIFFERENT LARGER STRUCTURES MICRO-UNIT CENTRAL INTELLIGENCE SYSTEM CENTERS AS WELL, THESE ARE CELL'S, CELL ORANS AND CELL SYSTEMS INDEPENDENT SEPARATE MICRO-UNIT'S CIS AND MICRO-UNIT'S PIS CENTERS. THE ALL INTERNAL CYTOPLASMIC AND INTERNAL NUCLEI CONSTRUCTIONS, WHICH THESE DIFFERENT INDEPENDENT MICRO-UNIT CIS ALL OPERATE UNDER THE ORDERS OF THE CELL'S (MICRO-UNITS) MAIN ESSENTIAL CENTERAL INTELLIGENCE SYSTEM CENTERS. WHICH CONTROLS INTER-CELL'S MICRO-UNIT SYSTEMS AND ORGANS INTELLEIGENCE SYSTEM CENTERS.

NANO-UNITS ALL POSSESSES INDIVIDUAL INDEPENDENT PARTICLE INTELLIGENCE SYSTEM CENTERS IN EACH GIVEN INTERNAL MICRO-UNIT CONSTRUCTION. AND THESE NANO – UNIT PIS STRUCTURES AS WELL AS INTERNAL CELL INDEPENDENT PARTICLE CIRCULATION SYSTEMS HELP ALL EXISTING NANO-UNITS INSIDE A GIVEN CELL OR INSIDE ANY GIVEN MICRO-UNIT, THE PIS ALL INDEPENDENTLY COMMUNICATE WITH EACH OTHER, UNDER CELL'S MAIN CELL'S ORIGINAL CENTERAL PARTICLE INTELLIGENCE SYSTEM CENTERS.

THERE ARE INTERNAL NANO- UNIT PARTICLE CIRCULATION SYSTEMS AND FUNDAMENTAL PARTICLE INFORMATION IMAGE PARTICLE CLOUD CIRCULATION SYSTEMS, WHICH CIRCULATE BETWEEN THE DIFFERENT NANO- UNIT PARTICLE INTELLIGENCE SYSTEM CENTERS (PIS), ALSO THERE ARE INTERNAL CELL'S OR INTERNAL MICRO-UNIT'S INDEPENDENT PIS (PARTICLE INTELLIGENCE SYSTEMS CENTERS) IN ALL MICRO – UNIT INTERNAL CELL NUCLEI, CYTOPLASMA, INTERNAL CHROMOSOMES, INTERNAL MITOCHONDRIA, AND ALL OTHER INTERNAL CELL SYSTEMS AND ORGANS PIS, WHICH ALL OF THESE ADDITIONALLY AND INDEPENDENTLY POSSESSES INDEPENDENT PARTICLE CIRCULATION SYSTEMS (PCS), WHICH ALL CONNECTS TO INDEPENDENT NANO-UNIT STRUCTURE ALL ONE BY ONE,

THE EXISTING INTERNAL CELL PARTICLE CIRCULATION SYSTEMS INSIDE CELL'S NUCLEI, CELL'S CYTOPLASMA, INTERNAL CELL DIFFERENT SYSTEMS AND INTERNAL CELL'S DIFFERENT ORGANS SUCH AS: CHROMOSOMES, MITOCHONDRIA, ENERGY PRODUCTION SYSTEMS, DUPLICATION SYSTEMS, ETC. AS WELL AS THE EXISTING INTERNAL NANO-UNIT PCS INSIDE ELECTRONS, NUCLEONS, ATOMS ALL CONNECT TO EACH OTHER, AND COMMUNICATE WITH EACH OTHER THROUGH THESE INTER-CONNECTED MICRO-UNITS COMPLEX PARTICLE CIRCULATION SYSTEMS, AND ALL THE INTERNAL CELL AND INTERNAL NANO-UNIT CENTRAL PARTICLE INTELLIGENCE THROUGH THESE PARTICLE INTELLIGENCE SYSTEM CENTERS, AND PARTICLE CIRCULATIONS SYSTEM CENTERS

CONNECT TO EACH OTHE AND DO PHYSICAL CHEMICAL AND BIOLOGICAL TASKS UNDER THE MAIN CELL'S PARTICLE INTELLIGENCE SYSTEM CENTER (CIS) ORDERS, WHICH IN THE SAME TIME THESE CENTERS ALL CONNECT TO THE ORDERS AND OPERATION OF THE OTHER SUPERIOR CELL'S ORDER SYSTEMS, OR INFERIOR CELL- POPULATION SYSTEMS ORDERS AND OPERATIONS, UNDER A STRICT HIERARCHY MASS ORDER SYSTEM (HERE THE MASSES ARE MICRO-UNITS MASS SYSTEMS AS WELL AS THE NANO-UNITS MASS SIZE OPERATION SYSTEMS WORKING TO GETHER AND PERFORMING BIOLOGICAL, CHEMICAL, AND PHYSICAL TASKS AT NANO AND MICO LEVELS).

PCS BETWEEN LIVING THINGS TOTAL BODY ELECTRONS NUCLEON POPULATIONS

1 - THE PARTICLE CIRCULATION SYSTEMS (P.C.S.) BETWEEN TOTAL BODY ATOMS, ELECTRONS, PROTONS, NEUTRONS, ETC. OF THE PLANET EARTH'S LIVING THINGS TOTAL – BODY NANO – UNIT POPULATIONS.

2 – THE INDEPENDENT PCS OF EACH GIVEN: ATOM, ELECTRON, NEUTRON, PROTON, AND NANO-UNITS IN ALL LIVE AND NON – LIVE NANO – UNIT POPULATIONS.

AND ENORMOUS RELATED DIFFERENT PARTICLE'S CENTRAL INTELLIGENCE SYSTEM CENTERS IN NANO-UNITS, MICRO-UNITS, AND MACRO-UNIT CONSTRUTCTIONS. ALL OF THE ABOVE DIFFERENT SUBJECTS ARE GOING TO BE EXPLAINED IN OTHER TEXTS SEPARATELY, HERE ONLY OUTLINES ARE PRESENTED ONLY AS:

CENTRAL INTELLIGENCE SYSTEM CENTERS OF EARTH'S EXISTING NANO-MICRO-MACRO MASS –UNITS:

================================

1- PIS OF THE NANO-UNITS

THE CENTRAL INTELLIGENCE SYSTEM CENTERS (C.I.S.) OF EACH GIVERN ATOM, ELECTRON, NUCLEON, AND OTHER UNIVERSAL LIVE OR NON – LIVE NANO – UNIT POPULATIONS.

2 - PIS OF THE MICRO – UNITS

THE CELL'S PARTICLE INTELLIGENCE SYSTEM CENTERS

C- PIS

THE CELL'S PIS COMPOSE FROM TWO DIFFERENT SEPARATE INDEPENDENT INTELLECTUAL OPERATION SYSTEMS CENTERS IN: CELL'S CYTOPLASMA PIS, AND CELL'S NUCLEI PIS, WHICH BOTH OF THESE PIS OPERATE UNDER CELL'S MAIN PIS CENTERS (M-PIS), ALL OF THESE OPERATE AND COORDINATE WITH EACH OTHER UNDER NANO-UNIT AND MICRO-UNIT MASS CONSTRUCTION LEVELS PHYSICAL CHEMICAL BIOLOGICAL OPERATIONS COOPERATIONS AS FOLLOWINGS:

1- NANO-UNIT LEVEL, FUNDAMENTAL PARTICLE OPERATED INTELLIGENCE SYSTEM CENTERS (FP- ISC), WHICH ARE INTELLIGENCE SYSTEM CENTERS OF INSIDE ELECTRONS, NUCLEONS, ATOMS AND CELL'S NANO- UNIT SYSTEM INTELLECTUAL SYSTEM CENTERS.

2– THE CELL'S MICRO-UNIT SIZE LEVELS MASSES CENTRAL PARTICLE INTELLIGENCE SYSTEM CENTERS (C - PIS) WHICH THE BIOMOLECULAR LARGER MASS PIS CAN BE CLASSIFIED UNDER THESE GROUPS AS WELL.

3- THE PIS BETWEEN THE DIFFERENT MICRO – UNITS SIZES OF THE DIFFERENT CELL'S ORGANS AND CELL'S SYSTEMS, AND LARGER CELL POPULATIONS DIFFERENT SPECIES MASS MICRO-UNITS PARTICLE INTELLIGENCE SYSTEM CENTERS (PIS). WHICH UNDER THESE PIS NUMEROUS DIFFERENT CELL'S SPECIES IN A GIVEN TISSUE – MASS UNITS, ARE CONTROLLED THROUGH THESE TISSUE PIS.

IT IS OBVIOUS THE DIFFERENT PARTICLE CIRCULATION SYSTEMS, EITHER INTERNAL ATOM, INTERNAL ELECTRON-NUCLEON'S PARTICLE CIRCULATION SYSTEMS (PCS), OR PARTICLE CIRCULATION SYSTEMS BETWEEN DIFFERENT INTERNAL CELL NUCLEI, INTERNAL CYTOPLASMIC DIFFERENT INTER-CELL ORGANS AND INTER-CELL'S SYSTEMS PARTICLE CIRCULATIONS SYSTEMS ALL CONNECT THROUGH THE PARTICLE CLOUD CURRENTS CIRCULATION SYSTEMS, WITH EACH OTHER, BETWEEN THE NANO-UNITS AND MAICRO-UNITS AS WELL AS MACRO-UNITS MASS – STRUCTURES.

THE INTERNAL CELL MASS -UNITS, AND OTHER NANO- MICRO-UNIT MASSES IN TISSUES ALSO CARRIED THROUGH THE COMMUNICATING FUNDAMENTAL PARTICLE INFORMATION IMAGE PARTICLE CLOUD CURRENTS OF THE DIFFERENT PCS, AND COMMUNICATE THROUGH THE EXCHANGES OF FP – I.I. – P. Cl. CIRCULATIONS SYSTEMS BETWEEN THE DIFFERENT PIS OF THE DIFFERENT MASS-UNIT SIZES LEVELS.

FOLLOWING ARE SOME EXAMPLE FROM DIFFERENT PARTICLE INTELLIGENCE SYSTEM CENTERS, WHICH CONNECT TO EACH OTHER THROUGH PARTICLE CIRCULATIONS SYSTEMS, THE DIFFERENT SIZES OF MASS UNITS COMMUNICATE WITH EACH OTHER THROUGH CIRCULATING FUNDAMENTAL PARTICLE CLOUDS, AND FP- I.I. – P. Cl. ETC. WITH EACH OTHER, WHICH PARTICLE CURRENTS CIRCULATE BETWEEN THE DIFFERENT MASS- SIZES PARTICLE INTELLIGENCE SYSTEM CENTERS THROUGH THE CIRCULATIONS OF THE PCS.

1 – THE CENTRAL INTELLIGENCE SYSTEM CENTERS (CIS) INSIDE DIFFERENT ELECTRON, NUCLEON, AND ATOM SYSTEM CENTER INSIDE CELL.

2 – CIS OF CELL'S CYTOPLASMA.

3 – CIS OF CELL'S NUCLEUS.

4 – CIS OF THE CELL.

5 - CIS INTERNAL CELL'S – ORGANS.

6 – CIS OF CYTOPLASMIC – ORGAN.

7 – CIS OF THE ORGANS IN CELL NUCLEUS.

8 – CIS OF THE INTERNAL CELL'S SYSTEMS,

THE CIS OF THE DIFFERENT INTERNAL NUCLEAR INTELLIGENCE SYSTEMS CENTERS, BIOMOLECULAR AND CHROMOSOMAL DUPLICATIONS SYSTEMS CENTERS, AND NUCLEAR DIVISION SYSTEMS CENTRAL PARTICLE INTELLIGENCE SYSTEM CENTERS (PIS), IN THE NUCLEUS AS WELL AS OTHERS IN CYTOPLASMA. OR CELL'S INTELLIGENCE SYSTEMS CENTERS FOR NUCLEI – TRANSPORTATION SYSTEMS IN NUCLEUS WHICH ARE SEPARATE THAN CYTOPLASMIC PIS OF THE CYTOPLASMIC PARTICLE CIRCULATION SYSTEMS AS WELL AS ATOM BASE MOLECULAR TRANSPORTATIONS SYSTEMS OF THE NUCLEI WHICH ARE DIFFERENT THAN CYTOPLASMIC TRANSPOTIONS SYSTEMS.

PIS FOR CELL'S IMMUNITY SYSTEMS CENTERS FOR BOTH NUCLEI AND CYTOPLASMA, WHICH ARE DIFFERENT FROM EACH OTHER, BUT CLOSELY COOPERATE AND COORDINATE THEIR FUNCTIONS WITH ACH OTHE AT ALL TIMES.

ABOVE FACT APPLY FOR OTHER INTERNAL CELL ORGANS AND OTHER INTERNAL CELL SYSTEMS AS WELL WITH NO EXCEPTIONS. FOR THE CELL'S DEFENSIVE AND IMMUNITY SYSTEMS CENTERS AS WELL.

ETC.

DIFFERENT LIVING THING SPECIES CIS CENTERS (CENTRAL INTELLIGENCE SYSTEM CENTERS) = CNS

1 - C.I.S. OF THE DIFFERENT LIVING THINGS BODY SYSTEM CENTERS

(S- CIS)

2 - C.I.S. OF DIFFERENT LIVING THINGS SPECIES OF BODY ORGAN'S INTELLECTUAL SYSTEMS CENTERS

(O- CIS)

1 – CIS OF THE LIVING THINGS BODY ORGANS.

2 – CIS OF THE LIVING THINGS BODY – SYSTEMS.

3 – CIS OF THE LIVING THINGS TOTAL BODY INTELLECTUAL SYSTEM CENTERS (CNS).

THE PRODUCED PARTICLE – CLOUDS, AND FUNDAMENTAL PARTICLE INFORMATION – IMAGE PARTICLE - CLOUD (F.P. – I.I. – P. Cl.) CURRENTS FROM DIFFERENT NON – LIVE AND LIVE CENTRAL INTELLIGENCE SYSTEM CENTERS (CIS) ATOMS, ELECTRONS, NUCLEON, AND OTHER NANO UNITS CONSTRUCTIONS OF ANY KIND ACROSS THE UNIVERCE, THROUGH DIFFERENT EXTERNAL OR INTERNAL BODY FUNDAMENTAL PARTICLE CIRCUIT CURRENTS AND PARTICLE CIRCULATION SYSTEMS (P.C.S.) CONNECT TO EACH OTHER BETWEEN THE TOTAL BODY NANO – UNIT POPULATIONS AT LIVING THING SPECIES.

THE DIFFERENT EXISTING ATOMS, ELECTRONS, NUCLEONS, AND NANO – UNITS IN ANY PLACE IN UNIVERCE, THROUGH THESE PARTICLE ROUTES, AND EXTERNAL BODY OR INTERNAL – BODY FUNDAMENTAL PARTICLE CIRCULATION SYSTEMS (EX. - PCS, & I- PCS) COMMUNICATE WITH EACH OTHER, THE UNIVERSAL DIFFERENT COMPATIBLE NANO – UNITS THROUGH THESE CLOSED INTERNAL OR EXTERNAL NANO – UNIT PCS, AND RPARTICLE CLOSED CIRCUIT CIRCULATIONS, THE FP – I.I.- P. Cl. UNDER A. S. I. FP. Mol. C.I.C. COMBINE WITH INTERNAL ELECTRON – NUCLEON PARTICLE COMPOUNDS OF ATOMS AND CONSTRUCT THE ELECTRONS, NUCLEON PARTICLE COMPOUND CONSTRUCTIONS, AND THIESE PHENOMENONS BRIEFLY ARE FUNDAMENTAL PARTICLE CIRCULATIONS SYSTEM (P.C.S.) BETWEEN THE DIFFERENT NANO – UNITS CENTRAL INTELLIGENCE SYSTEMS CENTERS (C. I. S.), AND BASICS OF ELECTRON –NUCLEON PARTICLE COMPOUND STRUCTURES CONSTRUCTIONS, STORAGE AND RETRIEVALS OF PARTICLE CLOUDS INSIDE ATOMS, AND NANO – UNIT –GENESIS PHENOMENONS IN UNIVERCE.

THE PARTICLE CLOUD CIRCULATION SYSTEMS (P. Cl. – PCS)

THE PCS EITHER TRAVEL THROUGH AIR AND ARE AIRBORN FUNDAMENTAL PARTICLE CLOUD PARALLEL BI-DIRECTIONAL CLOSED PARTICLE CIRCULATION SYSTEMS, OR THE PARTICLE CLOUDS TRAVEL THROUGH OTHER PATH FUNDAMENTAL PARTICLE TRANSIT CURRENTS AND TRANSMISSION ROUTES SUCH AS PARTICLE CIRCULATION SYSTEMS IN PERIPHERAL ATOM SPACES OF SOLIDS, LIQUIDS, PLASMA, ETC. WHICH THE PRODUCED S.Y. T.E.- F.P. – I.I.- P. Cl. OF ONE FP – ISC THROUGH PARTICLE CLOUD CIRCULATION SYSTEMS CURRENTS TRAVEL AND TRANSIT BETWEEN DIFFERENT NANO- UNITS FUNDAMENTAL PARTICLES INTELLECTUAL SYSTEM CENTERS, AND DIFFERENT NANO- UNITS CENTRAL INTELLIGENCE SYSTEM CENTERS OF COMMUNICATE WITH EACH OTHER AND EXCHANGE INFORMATION IMAGE PARTICLE CLOUDS WITH EACH OTHER, OR UNDER A. S. I. FP. Mol. C.I.C. THE PARTICLE CLOUDS OF ONE CENTER COMBINE WITH PARTICLE COMPOUNDS OF THE OTHER CISC OF THE ANOTHER ONES, AND STORE THE INFORMATIONS AND IMAGES OF ANOTHER CENTERS INSIDE RECIPIENTS ELECTRONS, NUCLEONS AND NANO-UNITS.

THESE ARE PHENOMENON OF NEO – INTELLIGENCE GENESIS STORAGE SYSTEMS, INSIDE THE DIFFERENT NANO- UNIT NEO-PARTICLE CLOUD COMPOUND CONSTRUCTIONS.

THE FUNDAMENTAL PARTICLE INTELLIGENCE SYSTEM CENTERS OF ELECTRONS, PROTONS, NEUTRONS

THE FUNDAMENTAL PARTICLES CIRCULATION SYSTEMS INSIDE ATOMS, ELECTRONS, NUCLEONS

THE NANO – UNITS INTELLIGENCE SYSTEM CENTERS (NU – PIS),

THE INTERNAL NANO – UNIT'S FUNDAMENTAL PARTICLES CIRCULATION SYSTEMS (PCS)

THE INTERNAL ATOM FUNDAMENTAL PARTICLES INTELLIGENCE SYSTEM CENTERS (PIS)

THE INTERNAL ATOMS FUNDAMENTAL PARTICLE CIRCULATION SYSTEMS,

ATOM'S FUNDAMENTAL PARTICLE CIRCULATION SYSTEMS

(A – PCS)

THE ENTIRE A- FP – CS ALL ARE BIDIRECTIONAL EFFERENT AND AFFERENT PARTICLE CURRENTS AND PARTICLE CIRCULATION SYSTEMS, SOME A – FP - CS ARE CENTRI-PETAL FUNDAMENAL PARTICLE CURRENTS, AND FUNDAMENTAL PARTICLE CIRCULATION SYSTEMS, AND THE OTHERS PARALLEL OPPOSITLY RUNNIG PARTICLE CURRENTS ARE CENTRI-FUGAL PARTICLE CURRENT CIRCULATIONS SYSTEMS, ALWAYS ONE GROUP PARTICLE CURRENTS RUNS PARRALEL IN OPPOSITE DIRECTIONS OF THE SECOND ONE.

THE ATOM'S FUNDAMENTAL PARTICLE CIRCULATION SYSTEMS DIVIDE INTO THREE MAJOR CLASSES:

1 - INTERNAL PERIPHERON FUNDAMENTAL PARTICLE CIRCULATION SYSTEMS (P – FP – CS),

2 – INTERNAL NUCLEUS FUNDAMENTAL PARTICLE CIRCULATION SYSTEMS (N – FP – CS),

3 – EXTERNAL ATOM FUNDAMENTAL PARTICLE CIRCULATION SYSTEMS, OR PERIPHERAL ATOM SPACE FUNDAMENTAL PARTICLE CIRCULATION SYSTEMS (PAS- FP - CS),

THESE A – FP – CS IN LIVING THINGS CONNECT THE TOTAL ORGANS AND SYSTEMS BODY ELECTRON – NUCLEON POPULATIONS OF PERIPHERAL BODY ATOM'S POPULATIONS, TO THE TOTAL CNS ELECTRON – NUCLEON'S CNS – TOTAL ATOM'S POPULATIONS, THROUGH THE ATOM'S, ELECTRON'S, NULEON'S INTER - FUNDAMENTAL PARTICLE CIRCULATION SYSTEMS.

AT THE SAME TIME, THE INTERNAL PERIPHERON FUNDAMENTAL PARTICLE CIRCULATIONS SYSTEMS CENTERS ADDITIONALLY SUB – DIVIDES INTO:

1 - INTERNAL ELECTRONS FUNDAMENTAL PARTICLE'S CIRCULATION SYSTEMS (E – FP – CS) WHICH CONNECTS THE ENTIRE PERIPHERAL ORGANS AND SYSTEMS TOTAL ELECTRON'S POPULATION NUMBERS ELECTRON'S – CENTRAL – INTELLIGENCE SYSTEM CENTERS (E – CISC) DIRECTLY TO THE CNS ELECTRONS, NUCLEONS AND ATOMS INTELLECTUAL CENTERS IN THE BRAIN, THROUGH THE DIFFERENT FUNDAMENTAL PARTICLE CIRCULATIONS SYSTEMS CURRENTS, AND INTERNAL ELECTRONS FUNDAMENTAL PARTICLES CIRCULATION SYSTEMS, IN CONSEQUENCE THE ENTIRE BODY ELECTRONS, NUCLEONS POPULATIONS, MUTUALLY CONNECT TO EACH OTHER THROUGH THESE DIFFERENT NUMEROUS FP – CIRCULATION SYSTEMS.

2 – THE EXTERNAL ELECTRON FUNDAMENTAL PARTICLE INTELLIGENCE SYSTEM CENTERS OF PERIPHERON CONNECT TO EACH OTHER THROUGH EXTERNAL ELECTRON FUNDAMENTAL PARTICLES CIRCULATIONS SYSTEMS OF PERIPHERON, (OR INTER – PERIPHERON FUNDAMENTAL PARTICLES CIRCULATION SYSTEMS AND FUNDAMENTAL PARTICLE INTELLIGENCE SYSTEM CENTER).

2 – THE INTERNAL NUCLEUS FUNDAMENTAL PARTICLE CIRCULATION SYSTEMS, AND INTER – NUCLEI INTELLIGENCE SYSTEM CENTERS ADDITIONALLY SUBDIVIDES INTO THREE SUBCLASSES:

1 – THE INTERNAL PROTON FUNDAMENTAL PARTICLE INTELLIGENCE SYSTEMS CENTERS, AND INTERNAL PROTON FUNDAMENTAL PARTICLES CIRCULATION SYSTEMS.

2 – THE INTERNAL NEUTRON FUNDAMENTAL PARTICLES INTELLIGENCE SYSTEMS CENTERS, AND INTERNAL NEUTRON FUNDAMENTAL PARTICLE CIRCULATIONS SYSTEMS.

3 – EXTERNAL NUCLEONS FUNDAMENTAL PARTICLES INTELLIGENCE SYSTEMS CENTERS, AND EXTERNAL NUCLEONS FUNDAMENTAL PARTICLES CIRCULATIONS SYSTEMS.

THE ENTIRE TOTAL BODY PROTON – NEUTRON, AND NUCLEUS POPULATIONS OF THE TOTAL BODY ORGANS AND SYSTEMS THROUGH THESE ELECTRON – NUCLEON, AND PERIPHERON – NUCLEI CENTRAL INTELLIGENCE SYSTEM CENTERS, AND THROUGH THESE INORMOUS FUNDAMENTAL PARTICLE CIRCULATIONS SYSTEMS CONNECT TO EACH OTHER, AND COORDINATE, COOPERATE THEIR FUNCTIONS WITH EACH OTHER, THROUGH FUNDAMENTAL PARTICLE PRECISION ACCURACIES WITH NO MISTAKES IN LIVING THINGS OF ALL DIFFERENT SPECIES ACROSS THE UNIVERCE, UNDER THE PHYSICAL, CHEMICAL, BIOLOGICAL LAWS OF THE FUNDAMENTAL PARTICLES.

FUNDAMENTAL PARTICLES (F.P.) OF NON- LIVING THINGS ADDITIONALLY OPERATE NON – LIVE ELECTRON – NUCLEON FUNCTIONS OF EXISTING THINGS ALL OVER UNIVERSE UNDER ABOVE RULES, WITH LESS COMPLEXICITIES, (NANO –NEUROLOGY OF LIVE AND NON- LIVE ELECTRONS – NUCLEONS IN UNIVERCE).

THE F.P. OPERATE THE FUNCTIONS OF NON LIVE

EXISTINGS THINGS ALL OVER UNIVERCE

===

HOW FUNDAMENTAL PARTICLES OPERATE LIVING THINGS

THE F.P. OPERATE FUNCTIONS OF THE LIVING THINGS ALL OVER UNIVERSE

THE F.P. CONSTRUCTED CENTRAL INTELLIGENCE SYSTEMS OF LIVING THINGS (C.I.S.)

THE F.P. CONSTRUCTED CENTRAL INTELLIGENCE SYSTEMS OF LIVING THINGS BODY SYSTEM CENTERS

THE F.P. CONSTRUCTED CENTRAL INTELLIGENCE SYSTEM CENTERS OF BODY ORGAN CENTERS

THE F.P. OPERATE FUNCTIONS OF THE LIVINGS BODY ORGANS, BODY SYSTEMS, AND LIVING THINGS AS A WHOLES THROUGH CONSTRUCTIONS OF THE F.P. CONSTRUCTED BODY INTELLIGENE SYSTEM CENTERS FOR LIVING THINGS, AND FOR LIVING THINS ORGANS AND SYSTEMS, AS WELL AS THE FUNDAMENTAL PARTCIES HAS CREATED FUNDAMENTAL PARTICLE CIRCULATION SYSTEMS AND PCS – NTELLECTUAL SYSTEMS CENTERS.

THE FUNCTIONS OF TOTAL BODY POPULATIONS OF LIVING THINGS ELECTRONS – NUCLEONS ALL CONTROLLED THROUGH LIVING THINGS DIFFERENTET C.I.S. – CENTERS OF TOTAL BODY FROM CNS, AS WELL AS INDIVIDUAL ORGAN AND SYSTEMS INDIVIDUAL CIS, WHICH EXISTED IN ALL ORGAND AND SYSTEMS OF THE LIVING THINGS.

BODY ORGANS ELECTRONS NUCLEONS POPULATIONS CONTROLLED BY ORDERS OF CENTRAL INTELLIGENCE SYSTEMS (CNS) ELECTRONS NUCLEON'S ORDERS.

ALL BODY SYSTEMS SUCH AS G.I.S., G.U.S., CARDIO-PULMONARY SYSTEMS, ETC. ALL HAVE SYSTEM SPECIFIC CENTRAL INTELLIGENCE SYSTEM CENTERS AS WELL AS SYSTEM SPECIFIC PARTICLE CIRCULATION SYSTEM, ABOVE FACTS ALSO ARE TRUE IN REGARD TO BODY ORGANS WHICH EACH GIVEN BODY ORGAN POSSESSES ORGAN SPECIFIC CENTRAL INTELLIGENCE SYSTEM AS WELL AS ORGAN SPECIFIC PARTICLE CIRCULATION SYSTEMS,THESE PARTICLE CIRCULATION SYSTEMS AND ORGAN CENTRAL INTELLIGENCE SYSTEMS MOSTLY ARE NANO PARTICLE CIRCULATION SYSTEMS AND NANO CENTRAL INTELLIGENCE SYSTEM CENTERS, PRESENTLY KNOWN ANATOMICAL OR HISTOLOGICAL NERVOUS SYSTEMS ARE NOT NANO STRUCTURES OR QUANTUM CONSTRUCTIONS.

CIRCULATION OF NORMAL PARTICLE CLOUDS WILL PRODUCE NORMAL FUNCTIONAL ORGANS AND SYSTEMS, IN THE CASES OF ABNORMALITY IN PARTICLE CLOUD CIRCULATIONS OR ABNORMAL F.P. – I.I. – P. cl. CIRCULATIONS WILL CAUSE FUNCTION DISORDERS AND IS THE CAUSE FOR LARGE NUMBERS OF IDIOPATHIC DISORDERS SUCH IRRITABLE G.I.S. DISORDERS, RESPIRATORY SYSTEMS AND G. U. S. DISORDER, NEUROSIS, ETC.

IN DIFFERENT LIVING THINGS SPECIES, EACH GIVEN BODY ORGAN CELL'S MOLECULES AND BIO-MOLECULES, LARGER INTERNAL CELL ORGANS AND SYSTEMS MOLECULAR AND BIOMOLECULAR ATOMS STRUCTURES (SUCH AS GENES AND CHROMOSOMALS MOLECULAR ATOMS ELECTRONS- NUCLEONS STRUCTUTRE AND CONSTRUCTIONS ALL FOLLOW THE RULES AND LAWS OF THE FUNDAMENTAL PARTICLES, WITH NO EXCEPTIONS.

EACH GIVEN BODY – ORGAN'S OF ANY GIVEN LIVING THING ELECTRON'S, NUCLEON'S AND ATOM'S, IN DIFFERENT INTERNAL CELL'S ATOM – BASE BIO- MOLECULAR AND ATOM- BASE MOLECULAR CONSTRUCTIONS, CONSTRUCT FROM THREE DISTINCT MAIN FUNDAMENTAL PARTICLE- COMPOUND CONSTRUCTION STRUCTURES:

1 – INTERNAL ELECTRON –NUCLEON FUNDAMENTAL PARTICLES INTELLIGENCE SYSTEM CENTERS, WHICH EXIST IN ALL TOTAL BODY ORGANS ELECTRONS – NUCLEONS AND NANO- UNITS POPULATIONS OF THE ALL BODY ORGANS CONNECTED TO EACH OTHER.

2- INTERNAL ELECTRON – NUCLEON FUNDAMENTAL PARTICLE CIRCULATION SYSTEM CENTERS, WHICH INDIVIDUALLY EXIST FOR ANY GIVEN NANO- UNITS FOR ALL BODY ORGANS AND SYSTEMS OF THE LIVING THINGS.

3 – NANO –UNIT STRUCTURAL FUNDAMENTAL PARTICLE – COMPOUND CONSTRUCTIONS, THAT CONSTRUCTING DIFFERENT ELECTRONS – NUCLEONS AND ATOMS CONSTRUCTIONS OF THE TOTAL BODY ORGANS, AND TOTAL BO SYSTEMS.

MALFUNCTION OF ANY INTERNAL NANO – UNIT CONSTRUCTION STRUCTURE CAN CAUSE MALFUNCTION OF RELATED BODY ORGANS AND ORGAN DISORDER.

THE ABOVE CONSTRUCTIONS ALL APPLY IN REGARD TO THE TOTAL BODY SYSTEMS AND ORGAN SIMILARLY WITH NO EXCEPTIONS.

THE ELECTRON – NUCLEON PARTICLE INTELLIGENCE SYSTEM CENTERS IN LIVING THINGS.

FUNDAMENTAL PARTICLE CIRCULATION SYSTEMS LIVING THINGS ATOMS, ACROSS THE UNIVERCE

1 – THE CENTRAL INTELLIGENCE SYSTEM CENTERS OF BODY ORGANS (O-CIS), AND INTER – BODY ORGANS PARTICLE CIRCULATION SYSTEMS.

THE CENTRAL INTELLIGENCE SYSTEM OF ORGANS ARE IN CHARGE OF ORGANS ELECTRONS NUCLEONS, PHYSICAL, CHEMICAL, BIOLOGICAL FUNCTIONAL OPERATIONS, AND MOST OF PHYSICAL CHEMICAL BIOLOGICAL FUNCTIONS OF ORGANS ELECTRON NUCLEON OPERATIONS CONTROLLED THROUGH BIOFRIENDLY PARTICLE CLOUDS AND ELECTRIC FUNDAMENTAL PARTICLE CLOUD ORDERS OF ORGANS CENTERAL INTELLIGENCE SYSTEMS OR THE PARTICLE CLOUD CURRENTS COMING FROM HIGHER OTHER INTELLIGENCE SYSTEM CENTERS AND PRODUCING EITHER NORMAL OR ABNORMAL FUNCTIONS ORGANS, CIS CURRENTS FLOWING THROUGH PARTICLE CIRCULATIONS SYSTEMS TO ENTIRE ELECTRON NUCLEON POPULATION OF ANY ORGAN, END RESULTS IS NORMAL OR ABNORMAL ORGAN FUNCTION.

FOR EXAMPLE, THE PARTICLE CLOUDS REPORTS OF NITROGLYCERIN (UNDER TONGUE) THROUGH ORGANS CIS - PCS TRANSIT TO CARDIAC ORGAN CAUSE VASODILATION AND PAIN RELIEF, AT THIS PHENOMENON ORAL NITROGLYCERIN'S INFORMATION PARTICLE CLOUD TRANSIT THROUGH LIGHT SPEED PARTICLE CLOUD CURRENTS AND PRODUCE SYMPTON CHANGE BY PARTICLE CLOUD INFORMATION PRODUCED VASODILATION.

BUT IN MANY CASES MALFUNCTION OF PARTICLE CIRCULATION SYSTEMS OR ABNORMALITY OF CIRCULATING PARTICLE INFORMATIONS AND IMAGE CLOUDS PRODUCE LARGE NUMBERS OF FUNCTIONAL DISORDERS IN DIFFERENT ORGANS OF MOST BODY SYSTEMS AND ORGANS WHICH ARE KNOWN PRESENTLY AS IDIOPATHIC, PARTICLE CLOUD ABNORMALITIES OR PARTICLE CIRCULATION SYSTEM ABNORMALITIES MAY PRODUCE SIMPLE ORGAN IRRITABILITY SUCH AS IRRITABLE BOWL SYNDROMME OR FUNCTION ABNORMALITY SUCH AS DYSPNEA, OR NEUROSIS, BUT LATER ADDED FURTHER COMPLICATIONS SUCH AS INFECTIONS OR INFLAMMATORY ORGANIC SECONDARY DISEASES FURTHER COMPLICATE AND CHANGE FOR ANOTHER DISEASE PRODUCTIONS LIKE CROHN'S, OR DESTRUCTIVE SECONDARY LUNG DISEASES.

THE ABNORMALITY IN PARTICLE CLOUDS, F.P.- I.I. – P.cl. AS WELL ABNORMALITY OF PARTICLE CIRCULATION SYSTEMS PRODUCE DISORDERS SUCH AS IRRITABLE BOWL, STOMACH, ETC. SYNDROMES, OR FUNCTIONAL DISORDERS IN BREATHINGS, SWALLOWING AND SIMILAR MANY OTHERS IN G. U. S. IRRITABLE DISORDERS WHICH AFTER PASSING TIMES MANY OTHER PERMANENT INFECTIONS OR INFLAMMATORY COMPLICATIONS ADD AND DEVELOPE FURTHER SERIOUS DISORDERS LIKE CROHN'S, ASTHMA, OR DESTRUCTIVE DISORDERS.

PRODUCTIONS OF ABNORMAL PARTICLE CLOUDS, ABNORMAL F.P. – I.I. – P.cl. BY CENTRAL INTELLIGENCE SYSTEM CENTERS, TRANSMISSION AND CIRCULATIONS OF THOSE ABNORMAL PARTICLE CLOUDS INSIDE ORGANS AND BODY SYSTEMS PARTICLE CIRCULATION SYSTEMS CIRCUITS PRODUCE ORGAN MALFUNCTIONS AND PRODUCE ABNORMAL SYMPTOMS AND SIGNS WHICH PATIENTS FEEL THOSE AND MOSTLY CALLED SOME THING IS WRONG IN BRAIN, IN

MANY CASES THERE IS MALFUNCTION AT PARTICLE CIRCULATIONS DIFFERENT TYPES WHICH ARE IN OTHER VOLUMES.

CENTRAL PARTICLE INTELLIGENCE SYSTEMS OF BODY SYSTEMS AND, FUNDAMENTAL PARTICLE CIRCULATION SYSTEM (P.C.S.) BETWEEN DIFFERENT BODY SYSTEMS.

FUNDAMENTAL PARTICLE CIRCULATION SYSTEMS OF BODY ORGANS MOSTLY ARE FLOWING ELECTRIC FUNDAMENTAL PARTICLES CURRENTS AND PARTICLE CLOUD CURRENTS OF ELECTRIC PARTICLES AS WELL OTHER CLASS FUNDAMENTAL PARTICLE CLOUD CURRENTS SUCH AS LIGHT PARTICLE CLOUDS , SONIC PARTICLE CLOUDS, THERMAL PARTICLES CURRENTS, ETC., WHICH THESE PARTICLE CLOUDS PRODUCED AND ORIGINATED FROM CENTRAL INTELLIGENCE SYSTEM CENTERS, AND FROM THERE FLOW INSIDE NANO-NEURAL FIBERS PARTICLE CIRCULATION SYSTEMS DUCTS AS PARTICLE CURRENTS FINALLY INTERS INTO INTERNAL ELECTRONS - NUCLEONS POPULATIONS OF DIFFERENT ORGANS.

3 – the TOTAL BODY ELECTRON NUCLEON POPULATIONS, AND TOTAL BODY INTERNAL ELECTRON – NUCLEON PARTICLE CIRCULATION SYSTEMS.

The CENTRAL INTELLIGENCE SYSTEM CENTERS FOR TOTAL BODY ELECTRONS – NUCLEONS POPULATIONS.

NANO NEUROLOGY OF THE TOTAL BODY SYSTEMS:

NANO FUNCTIONS OF TOTAL ELECTRONS NUCLEONS POPULATIONS IN A BODY SYSTEM, IS UNDER NANO INTELLIGENCE CENTERS OF BODY SYSTEMS

FROM NANO-NEUROLOGICAL POINT OF VIEW EACH GIVEN BODY SYSTEM CONSTRUCTED FROM: 1 – C. I. S., 2 – P. C. S., 3 - NANO-UNIT STRUCTURES OF BODY SYSTEMS:

CENTRAL INTELLIGENCE CENTERS OF BODY SYSTEMS,

CENTRAL INTELLIGENCE CENTERS ORDER IN PARTICLE CLOUD FORMS, TRANSIT PARTICLE CLOUDS THROUGH PARTICLE CIRCULATIONS SYSTEMS TO DIFFERENT ORGANS ELECTRON NUCLEONS POPULATIONS, PARTICLE CLOUDS INTER INTO ELECTRONS, NUCLEONS AND UNDER A.S. I. F.P. Mol. C.I.C. INTERACT WITH EACH OTHER, ALL EXISTING ORGANS IN A GIVEN BODY SYSTEM FUNCTION UNDER BODY SYSTEMS CENTRAL INTELLIGENCE CENTERS ORDERS PARTICLE CLOUD INSTRUCTIONS,

THE BODY SYSTEM'S CENTRAL INTELLIGENCE CENTERS PRODUCE AND REPORT F.P. – I.I. – P. cl. (PARTICLE CLOUDS) TO HIGHER CNS AND CENTRAL AUTONOMOUS SYSTEMS CENTERS AND FOLLOW THESE CENTERS ORDERS AND INSTRUCTIONS, THE C. I. S. ALSO ARE CONTROLLED THROUGH END NANO-UNIT FINDING REPORTS AND RECOMMENDATIONS, ALTHOUGH ARE ISSUERS OF ORDERS TO LOWER INTELLIGENCE CENTERS,

DIFFERENT BODY SYSTEMS MOSTLY THROUGH ADJACENT ADDITIONAL SIDE PARTICLE CIRCULATION SYSTEMS, CONNECT TO OTHER NEIGHBORING ORGANS AND SYSTEMS, THROUGH THESE SIDE PARTICLE CIRCULATION SYSTEMS CONNECTIONS THE SYMPTOMS AND SIGNS OF ONE SYSTEMS OR ONE ORGAN TRANSMIT AND SENSED AT NEIGHBORING SYSTEM OR ORGANS IN MANY EPISODES,

PLANT'S AND ANIMAL'S NANO NEUROLOGY

PHENOMENON OF COMMUNICATIONS OF TOTAL BODY ELECTRONS NUCLEONS POPULATION WITH EACH OTHER BY PARTICLE CLOUDS.

F.P.- I.I. P. cl. EXCHANGE, AND PARTICLES CIRCULATION SYSTEMS BETWEEN TOTAL - BODY ELECTRONS NUCLEONS POPULATIONS.

HIERARCHY PHENOMENONS BETWEEN ELECTRONS NUCLEONS SOCIETY, THROUGH PARTICLE CLOUD EXCHANGE ORDERS.

HIERARCHY CELL SOCIETY SYSTEMS PHENOMENONS IN CELLS COMMUNICATION SYSTEMS

PHENOMENON OF COMMUNICATION BETWEEN TOTAL BODY ELECTRONS NUCLEONS, THROUGH PARTICLE CLOUD EXCHANGE SYSTEMS BETWEEN ELECTRONS NUCLEONS, AND FUNDAMENTAL PARTICLES CLOUD CIRCULATION SYSTEMS (P C S) THROUGH NANO NEURAL DUCT CIRCULATION SYSTEMS IN ANIMALS DIVIDE INTO THREE MAJOR PARTICLE CIRCULATION SYSTEM (P. C. S.) CATEGORIES CLASSIFICATIONS:

1 – INDIGENOUS INTER - ELECTRON - NUCLEON F.P. –I.I.- P.cl. EXCHANGE, HIERARCHY SYSTEMS AND PARTICLE CIRCULATION SYSTEMS (I - PCS).

2- EXOGENOUS INTER –ELECTRON-NUCLEON PARTICLE CLOUD EXCHANGE, AND EXOGENOUS PARTCLE CIRCULATIONS SYSTEMS (EX - PCS).

3- CONNECTIONS AND PARTICLE CIRCULATION EXCHANGES BETWEEN: 1 & 2 (I - PCS & EX - PCS.).

TOTAL BODY PARTICLE CIRCULATING SYSTEMS (TB – PCS)

THIS IS PCS BETWEEN TOTAL BODY ELECTRONS, NUCLEONS POPULATION OF ANIMALS BODY, THIS IS THE FUNDAMENTAL PARTICLE CIRCULATION SYSTEMS THAT, CIRCULATE BETWEEN WHOLE-BODY ELECTRONS, NUCLEONS POPULATIONS INSIDE ANIMAL BODY, CONNECTS ALL BODY'S ATOMS TO EACH OTHER.

INDIGENOUS PARTICLE CIRCULATION SYSTEMS

1 - INDIGENOUS PARTICLE CLOUD CIRCULATION SYSTEMS INSIDE ANIMAL BODY BETWEEN ELECTRONS NUCLEONS, AND INDIGENOUS INTER -ELECTRON - NUCLEON PARTICLE CLOUD EXCHANGES UNDER HIERARCHY SYSTEMS, THROUGH PARTICLE CLOUD EXCHANGE.

IN ANIMALS SPECIES, THE INDIGENOUS PARTICLE CIRCULATION SYSTEMS (I. PCS) ARE BI-DIRECTIONAL PARTICLE CLOUD TRANSPORTATION SYSTEMS, THE PCS CIRCULATE PARTICLE CLOUDS AND F.P. – I.I.- P. cl. BETWEEN TOTAL BODY ELECTRONS –NUCLEONS POPULATIONS UNDER HIERARCHY ORDER AND RESPOSE PARTICLE CLOUD CIRCULATION SYSTEMS.

THE PARTICLE CLOUD CIRCULATION SYSTEMS (PCS) TRANSMIT PERIPHERAL ELECTRONS NUCLEONS PARTICLE CLOUD INFORMATION AND IMAGES CURRENTS FROM PERIPHERAL ORGANS ELECTRONS NUCLEONS THROUGH NANO FIBER NEURAL DUCTS AND CARRY THOSE PARTICLE CLOUDS TO HIERARCHY HIGHER INTELLECTUAL CENTERS OF CNS AND AUTONOMOUS CENTER SYSTEMS ELECTRONS NUCLEONS REVIEWS AND DECISIONS.

THE HIGHER INTELLIGENCE SYSTEM CENTERS ELECTRONS NUCLEONS THROUGH DIFFERENT CENTER REVIEW A. S. I. F.P. Mol. C.I.C. PRODUCE ORDER - PARTICLE CLOUDS, THE ORDER PARTICLE CLOUDS THROUGH REVERSE DIRECTION PARTICLE CIRCULATION SYSTEM CURRENTS GET TRANSPORTIONS BACK TO PERIPHERAL ORGANS ELECTRONS NUCLEONS POPULATIONS AS INSTRUCTIONS FOR FUNCTION, WHICH UNDER THESE ORDER PARTICLE CLOUDS THE PERIPHERAL ORGAN ELECTRONS NUCLEONS FUNCTION ACCORDINGLY AND PERFORM NEEDED PHYSICAL CHEMICAL BIOLOGICAL NANO FUNCTIONS, ABOVE IS ONE CYCLE PARTICLE CIRCULATION SYSTEMS CYCLE WHICH TRANSPORT PARTICLE CLOUDS BETWEEN DIFFERENT LEVEL ELECTRONS NUCLEONS HIERARCHY PARTICLE CLOUDS ORDERS AND RESPONSES SYSTEMS.

EACH GIVEN BODY ORGAN AND BODY SYSTEM POSSESSES ITS ORGAN SPECIFIC CENTRAL INTELLIGENCE SYSTEM CENTERS (C. I. S.), AS WELL AS SYSTEM SPECIFIC CENTRAL INTELLIGENCE SYSTEM CENTERS (C. I. S.), WHICH ALL OF THESE INDEPENDENT ORGANS C.I. S. AND BODY SYSTEMS C. I. S. ORDER THE TOTAL BODY POPULATION ELECTRONS AND NUCLEONS HOW THEY MUST PERFORM THE INTERNAL ELECTRON NUCLEON CHEMICAL, BIOLOGICAL AND PHYSICAL FUNCTIONS AND A. S. I. F.P. Mol. C.I.C. OF TOTAL BODY ORGANS AND SYSTEM.

THE TOTAL BODY ELECTRON NUCLEON POPULATION MEAN THE ALL ENTIRE ELECTRON-NUCLEON POPULATIONS THAT EXIST INSIDE THE BODY ORGANS AND BODY SYSTEMS OF A GIVEN ANIMAL SPECIES,

2 - EXOGENOUS PARTICLE CIRCULATION SYSTEMS (EX – PCS)

FUNDAMENTAL PARTICLE CIRCULATION SYSTEMS BETWEEN TWO DIFFERENT ANIMALS CENTRAL INTELLIGENCE SYSTEMS ELECTRONS NUCLEONS,

AIRBORN F.P. – I.I. – P. cl. EXCHAGE PHENOMENON BETWEEN TWO OR MORE ANIMALS CNS ELECTRONS NUCLEON POPULATIONS, THROUGH EX. PCS

THE AIRBORN EXOGENOUS PARTICLE CLOUD CIRCULATION SYSTEMS,

IN EXOGENOUS PARTICLE CIRCULATION SYSTEMS (EX – PCS) BETWEEN TWO OR MORE DIFFERENT INDIVIDUALS EXCHANGE THEIR CNS PRODUCED F.P. – I.I. – P. cl. BETWEEN EACH OTHER THROUGH AIRBORN PARTICLE CIRCULATION SYSTEMS, THE EXOGENOUS PARTICLE CLOUD CIRCULATION SYSTEMS (EX-PCS) IN AIR TRAVEL AT MULTIPLE DIRECTIONS BETWEEN MULTI-INDIVIDUALS THROUGH CLOSED PARTICLE CLOUD CIRCULATION CYCLES,

EACH INDIVIDUAL'S CNS ELECTRONS NUCLEONS PRODUCE DIFFERENT PARTICLE CLOUDS, F.P. – I.I.- P.cl. AND EMIT PARTICLE CLOUDS AIRBORN, THE AIRBORN PARTICLE CLOUDS TRAVEL IN ALL DIRECTIONS IN AIR, THESE AIRBORN PARTICLE CLOUDS ARE CAPTURED BY DIFFERENT MULTIPLE INDIVIDUAL RECIPIENTS CNS ELECTRONS NUCLEONS, IN RECIPIENTS THE PARTICLE CLOUDS INTER INTO RECIPIENT INDIVIDUALS CNS ELECTRONS NUCLEONS, COMBINE WITH INTER ELECTRON NUCLEON PARTICLE COMPOUNDS, THESE PARTICLE CLOUD COMMPOUNDS AND EXOGENOUS PARTICLE CLOUDS INTERACT UNDER A. S. I. F.P. Mol. CIC BETWEEN DIFFERENT CNS CENTERS ELECTRONS NUCLEONS, FINALLY RECIPIENT ELECTRONS NUCLEONS UNDER INNORMOUS INTERACTIONS PRODUCE FUNDAMENTAL PARTICLE RESPONSE DIRECTLY RELATED TO DONNER'S INCOMING AIRBORN PARTICLE CLOUDS, THE PRODUCED RESPONSE PARTICLE CLOUDS EMIT INTO SPACE AND AIRBORN TRAVEL IN MULTI- AIR DIRECTIONS AND FINALLY CAPTURED BY RECIPIENTS CNS ELECTRONS NUCLEONS POPULATIONS FOR INTERACTIONS, THIS IS ONE CYCLE AIRBORN PARTICLE CLOUD CIRCULATIONS SYSTEMS WHICH CIRCULATE BETWEEN DIFFERENT RECIPIENTS AND PARTICLE CLOUD DONNER'S ELECTARONS NUCLEONS THROUGH PARTICLE CLOUD CIRCULATION SYSTEMS, THESE ARE PHENOMENONS OF AIRBORN MULTI- EXOGENOUS PARTICLE CLOUD CIRCULATIONS SYSTEM WHICH PRESENTLY CALLED COMMUNICATIONS.

3 - PHENOMENON OF DIRECT CONNECTIONS BETWEEN: AIROGENOUS P.C.S. (EXOGENOUS EXTERNAL – BODY PARTICLE CIRCULATION SYSTEMS) WITH INTERNAL BODY INDIGENOUS PARTICLE CIRCULATION SYSTEMS, AND CONTINUATIONS OF THESE TWO PCS ALWAYS CONNECTED TO EACH OTHER AS A SINGLE CIRCUIT P.C.S.

THE EXOGENOUS PARTICLE CIRCULATION SYSTEM'S INTERCONNECTION WITH INDOGENOUS PARTICLE CIRCULATION SYSTEMS CIRCUITS SYSTEMS

HOW INDIGENOUS F.P.- I.I.- P.cl. –GENESIS INTERCONNECT AND INTERACT WITH EXOGENOUS PARTICLE CLOUD – GENESIS?

HOW CNS ELECTRONS NUCLEONS RESPOND TO INCOMING PARTICLE CLOUDS OF OTHER INDIVIDUALS CNS ELECTRONS NUCLEONS PARTICLE CLOUDS?

THE PHENOMENON OF INTERACTIONS BETWEEN EXOGENOUS PARTICLE CLOUD CIRCULATION SYSTEMS (EX-PCS) PARTICLE CLOUDS, WHEN THEY ARE INTERACTING WITH INDIGENOUS PARTICLE CLOUD CIRCULATION SYSTEMS (I-PCS) BRIEFLY EXPLAINED IN PREVIOUS PAGES.

DURING INTER INDIVIDUAL COMMUNICATIONS BETWEEN TWO OR MORE INDIVIDUALS, THE DONNER'S CNS PRODUCED F.P. –I.I.- P. cl. UNDER A.S. I. F.P. Mol. C.I.C. EMIT FROM DONNER'S ELECTRONS NUCLEONS AND LEAVES OUT OF CNS ATOMS AS AIRBORN FREE PARTICLE CLOUDS IN SPACE, AND TRAVEL DIFFERENT DIRECTIONS.

THE RECIPIENT INDIVIDUALS CNS ELECTRONS NUCLEONS CAPTURE THESE AIRBORN FLOATING PARTICLE CLOUDS, AND ANALIZE THE F.P. – I.I. – P.cl. UNDER A. S. I. F.P. Mol. CIC IN RELATED DIFFERENT CNS CENTERS AND PRODUCE RESPONSE – F.P. – I.I. – p. cl. OF RECIPIENT, AND RELEASE RESPONSE PARTICLE CLOUDS AIRBORN TO SPACE, UNDER SAME CYCLES OF PARTICLE CIRCULATION SYSTEMS THE DONNER CNS ELECTRONS NUCLEONS CAPTURE FLOATING AIRBORN PARTICLE CLOUDS OF RECIPIENTS, THESE PHENOMENONS CONTINUES BACK AND FORTH IN CLOSED CYCLES OF ALTERNATING INDIGENOUS PARTICLE CLOUDS CIRCULATION SYSTEMS ALTERNATE WITH EXOGENOUS PARTICLE CLOUD CIRCULATION SYSTEMS BACK AND FORTH, THIS IS BRIEFLY PHENOMENON OF COMMUNICATIONS BETWEEN DIFFERENT INDIVIDUALS THROUGH PARTICLE CLOUD EXCHANGE SYSTEMS.

ABOVE EXPLAINED PORCESS IS SUM OF ONE EXOGENOUS AIRBORN EXTERNAL BODY PARTICLE CLOUD CIRCULATION SYSTEMS (EX – PCS) CYCLE PHENOMENONS INTERACTIONS ALTERNATING WITH INDIGENOUS PARTICLE CLOUD CIRCULATION SYSTEMS, THIS IS PHENOMENON OF EX – PCS INTERACTIONS WITH I – PCS.

PIS OF BODY ORAGANS, AND PIS OF BODY SYSTEMS

EACH GIVEN ANIMAL BODY ORGAN, POSSESSES INDEPENDANT CENTRAL ORGAN'S PARTICLE INTELLIGENCE SYSTEMS CENTER (O - PIS).

EACH GIVEN ANIMAL BODY SYSTEM, POSSESSES INDEPENDANT BODY SYSTEM'S CENTRAL PARTICLE INTELLIGENCE SYSTEM CENTERS (S- PIS).

ALL BODY ELECTRONS NUCLEON POPULATIONS FUNCTIONS, UNDER PARTICLE CLOUD ORDERS, THAT HAS BEEN ISSUED BY UPPER PARTICLE INTELLIGENCE SYSTEMS ORDERS, FUNCTIONS ACCORDINGLY, UNDER THE ISSUED ORDERS, ALSO RESPOND BACK TO THOSE INCOMING CENTRAL PARTICLE INTELLIGENCE SYSTEMS.

IN LIVING THINGS PARTICLE CLOUD CURRRENTS CIRCULATE BETWEEN DIFFERENT FUNDAMENTAL PARTICLE INTELLIGENCE SYSTEM CENTERS ALSO CONNECT ALL BODY INTELLIGENCE SYSTEMS AND ORGANS INTELLIGENCE SYSTEMS CENTERS WITH TOTAL ELECTRON NUCLEON POPULATIONS IN ANIMAL BODY, ALSO PARTICLE CIRCULATION SYSTEMS CARRY AND TRANSIT DIFFERENT NEEDED FUNDAMENTAL PARTICLES FROM PARTICLE DONNER LOCATIONS ELECTRONS NUCLEONS TO THE RECIPIENT INTERNAL ELECTRON NUCLEON SYSTEMS, AND REMOVE THE WASTE AND NOT NEEDED PATICLES OUT OF DIFFERENT NANO- LOCATIONS OF LIVING THINGS BODY, TWO OPPOSITE DIRECTION DIFFERENT AFFERENT AND EFFERENT PARTICLE CURRENTS AND FUNDAMENTAL

PARTICLE CLOUD CIRCULATION SYSTEMS CONNECT ALL TOTAL BODY ELECTRON NUCLEON POPULATIONS TO EACH OTHER.

INDEPENDANT INTELLIGENCE SYSTEMS OF BODY ORGANS AND BODY SYSTEMS ELECTRONS AND NUCLEONS THROUGH INTER ELECTRON NUCLEON A. S. I. F.P. Mol. C.I.C. PRODUCE INSTRUCTIONS AND ORDER PARTICLE CLOUDS, AND THESE PARTICLE CLOUDS THROUGH NANO-NEURAL PARTICLE CLOUD CIRCULATION SYSTEMS PARTICLE CLOUD CURRENTS, TRANSPORT TO ALL PERIPHERAL BODY ELECTRONS NUCLEONS POPULATIONS.

THE CENTRAL INTELLIGENCE SYSTEMS ISSUED PARTICLE CLOUDS ORDERS, INTER INTO BODY ORGANS PERIPHERAL ELECTRONS NUCLEONS SUBSYSTEM –UNITS, AND DIFFERENT BODY SYSTEMS AND ORGANS ELECTRONS NUCLEONS UNDER THOSE PARTICLE CLOUD ORDERS ACHIEVE PHYSIOLOGICAL, CHEMICAL AND BIOLOGICAL FUNCTIONS OF LIVING THINGS ACCORDING TO THE INSTRUCTIONS OF HIGHER INTELLIGENCE SYSTEM CENTERS ELECTRONS AND NUCLEONS PARTICLE CLOUD ORDERS.

THE TOTAL ELECTRON NUCLEON POPULATIONS OF LIVING THINGS BODY, THE TOTAL ELECTRON- NUCLEON POPULATION OF BRAIN, THE TOTAL ELECTRON NUCLEON POPULATIONS OF BODY ORGANS AND BODY SYSTEMS OF LIVING THINGS VARIES REMARKABLY FROM EACH OTHER.

PHENOMENON OF PSYCHE – GENESIS

AND PHENOMENON OF THOUGHT CURRENT – GENESIS IN LIVING THINGS

PHENOMENON OF S.Y.– FP– I.I. – P.cl. STORAGE INSIDE CNS NANO-UNITS

(STORAGE OF KNOWLEDGE)

AIRBORN EXOGENOUS INCIDENT BIOFRIENDLY ELECTRIC FUNDAMENTAL PARTICLES, LIGHT FUNDAMENTAL PARTICLE INFORMATION IMAGE PARTICLE CLOUDS (Y. F.P. – I.I.- P. cl.), SONIC FUNDAMENTAL PARTICLE INFORMATION IMAGE PARTICLE CLOUDS (S. F.P. – I.I.- P.cl.) INTER INTO CNS INTERNAL ELECTRON NUCLEON CHEMICAL LAB.S AND COMBINE WITH INDOGENOUS PRE-EXISTING INTERNAL ELECTRON-NUCLEON PARTICLE – COMPOUNDS AND CONSTRUCT CNS ELECTRONS-NUCLEONS PARTICLE COMPOUNDS CONSTRUCTIONS UNDER REGENERATIVE A.S. I. F. P. Mol. C.I.C.

THIS IS PHENOMENON OF NEO- S. Y. – F.P. –I.I. – P.cl. _ COMPOUND CONSTRUCTION (NEO-GENESIS OF PARTICLE – CLOUD COMPOUNDS) OF THE CNS INTERNAL ELECTRONS NUCLEONS CONSTRUCTIONS, AS WELL AS THIS IS PHENOMENON OF STORING SONIC – PARTICLE INFORMATION – IMAGE PARTICLE CLOUDS, AND LIGHT FUNDAMENTAL PARTICLE INFORMATION IMAGE PARTICLE CLOUDS INSIDE THE CNS ELECTRONS AND NUCLEONS, IN THE CONSTRUCTION FORMS OF LIGHT- SONIC PARTICLE CLOUD –COMPOUND CONSTRUCTIONS (S. Y. – F.P. – I.I.- P cl. _ COMP. - GENESIS),

DURING LIFE, ANIMALS COMPILE STORAGE OF DIFFERENT ENVIRONMENTAL SUBJECTS PARTICLE –CLOUDS INFORMATIONS AND IMAGES (S. Y. – F.P. – I.I. – P.cl.) IN PARTICLE COMPOUND FORMS, ALL PARTICLE CLOUDS ARE STORED INSIDE CNS ELECTRONS –NUCLEONS, THESE INFORMATIONS AND IMAGES PARTICLE COMPOUND STORAGES INSIDE CNS ELECTRONS-NUCLEONS ARE STORAGE OF KNOWLEDGE AVAILABLE FOR RETRIEVAL TO OUT OF CNS ELECTRONS NUCLEONS ALL TIMES IN ANY MOMENTS.

THESE STORED PARTILE CLOUD COMPOUNDS ARE BUILDING BLOCKS OF KNOWLEDGE AND ARE USED IN LEARNING EDUCATIONS, TRAININGS FOR INDIVDUALS, AND INTELLIGENCE CONSTRUCTIONS INSIDE CNS ELECTRONS NUCLEONS OF LIVING THINGS CNS AS WELL AS NON-LIVE ELECTRONS- NUCLEONS PARTICLE COMPOUND CONSTRUCTIONS, THESE STORED INFORMATIONS AND IMAGES ARE AVAILABLE FOR RETRIEVAL TO OUTSIDE PARTICLE-CLOUD –COMPOUND FORMS TO OUTSIDE ELECTRONS AND NUCLEONS.

THIS IS PHENOMENON OF PSYCHE-GENESIS AND STORAGE OF SCIENCE AND INFORMATION IMAGE INSIDE ELECTRONS NUCLEONS IN PARTICLE COMPOUND FORMS, INTERACTIONS OF THESE SONIC LIGHT FUNDAMENTAL

PARTICLE COMPOUND INFORMATION IMAGE PARTICLE CLOUDS WITH EACH OTHER AS WELL AS WITH EXOGENOUS INCOMING S. Y. – F.P. – I.I. – P. cl. PRODUCE THOUGHT CURRENTS.

THE THOUGHT CURRENTS ARE SONIC LIGHT FUNDAMENTAL PARTICLE INFORMATION IMAGE PARTICLE CLOUD CURRENTS AND INTERACTIONS OF THESE PARTICLE CLOUD CURRENTS WITH EACH OTHER INSIDE CNS ELECTRONS NUCLEONS PRODUCE PSYCHE, BIOLOGICAL SENSING OF PARTICLE CLOUD CURRENTS SENSE AS THOUGHT CURRENTS, THIS IS PHENOMENON OF PSYCHE –GENESIS AND THOUGHT CURRENT-GENESIS.

THE ATTRACTION OF S.Y. –F.P. – I.I. – P. cl., AND OTHER PARTICLE CLOUDS FROM AIR

PHENOMENON OF PSYCHE-GENESIS AND THOUGHT CURRENT –GENESIS

THE ATTRACTIVE GRAVITON FORCES OF SENSARY ORGANS ELECTRONS-NUCLEONS ATTRACT S. Y. – F.P. – I.I. – P. cl. FROM AIR, PARTICLE CLOUDS INTER INTO SENSARY ORGANS ELECTRONS AND NUCLEONS, THE INCIDENT S. –F.P. – I.I. – P.cl., Y. - F.P.- I.I. – P.cl., T. – F.P. – I.I. – P. cl., ETC., THROUGH PARTICLE CLOUD NEURAL NANO –DUCTS PARTICLE CURRENT CIRCULATIONS SYSTEMS, INTER INTO DIFFERENT CNS SENSARY –CENTERS INTERNAL ELECTRONS-NUCLEONS SUBSYSTEM UNITS CHEMICAL LAB.S, AND COMBINE WITH INTER CNS INTERNAL ELECRTON NUCLEON PARTICLE COMPOUNDS, UNDER REGENERATIVE A. S. I. F.P. Mol. C.I.C., AND PRODUCE F.P. – I.I. – P. cl. _ COMPOUNDS, AND CONSTRUCT DIFFERENT CNS CENTERS NEW ELECTRONS NUCLEONS PARTICLE CLOUD MOLECULAR CONSTRUCTIONS,THIS PHENOMENON IS PARTICLE –CLOUD STORAGE PHENOMENON INSIDE CNS DIFFERENT SENSARY CENTERS ELECTRONS –NUCLEONS IN THE FORMS OF PARTICLE CLOUD COMPOUNDS.

THE REVERSE OF ABOVE INTERACTIONS UNDER DEGENERATIVE A. S. I. F.P. Mol. C.I.C. CAUSE BREAK DOWN OF LARGER PARTICLE CLOUD COMBOUNDS INTO CONSTRUCTING SMALLER ORIGINAL PARTICLE CLOUD STRUCTURES, THESE AUTONOMOUS SEQUENTIAL INTERACTIONS AND PRODUCED SONIC- LIGHT PARTICLE CLOUD CURRENTS AND PARTICLES INTERACTIONS WITH EACH OTHER SENSE AS THOUGHT CURRENTS, PSYCHE AND THOUGHT CURRENTS SYSTEMS.

RELEASE OF LIGHT SONIC PARTICLE CLOUDS FREE FROM S. Y. – F.P. – I.I. – P. cl. _ COMPOUNDS FORMS FOLLOWING BREAK DOWN OF PARTICLE COMPOUNDS UNDER DEGENERATIVE A. S. I. F.P. Mol. C.I.C. PRODUCE FREE SONIC LIGHT PARTICLE CLOUDS CURRENTS, AND THIS PHENOMENON IS RETRIEVAL AND RELEASE OF PARTICLE CLOUDS TO OUT OF PARTICLE CLOUD COMPOUND FORMS AND PRODUCTIONS FREE PARTICLE CLOUD CURRENTS INSIDE CNS ELECTRONS NUCLEONS CIRCULATING BETWEEN DIFFERENT CNS CENTERS, WHICH IT IS THOUGHT CURRENT GENESIS AND PSYCHE GENESIS.

ATOM'S TURN OVER PROCESS, EQUALS TO EVOLUTION PATHS OF ATOM GENESIS

DEFINITION OF TURN OVER PHENOMENON:

PERIODICALLY DURING EACH ONE HALF LIFE OF ATOM, ELECTRONS, NUCLEON, THE ENTIRE PRE-EXISTING PARTICLE COMPOUNDS CONSTRUCTIONS OF DIFFERENT ELECTRONS, NUCLEONS, AND ATOM ALL REMOVED, AND IN THEIR PLACE ANOTHER EXACT COPY NEW EXACTLY THE SAME PARTICLE COMPOUNDS RECONSTRUCTED AS BEFORE, THIS IS PHENOMENONS OF THE ATONS, ELECTRONS, NUCLEONS TURN OVER PHENOMENO.

THE DESTRUCTIONS OF THE ELECTRONS, NUCLEONS, AND ATOMS PARTICLE COMPOUNDS ARE DONE THROUGH DEGENERATIVE A. S. I. P.F. Mol. C.I.C., AND THE REGENESIS OF ELECTRONS, NULEONS AND ATOMS PARTICLE COMPOUND CONSTRUCTIONS, TAKE PLACE THROUGH REGENERATIVE AUTONOMOUS SEQUENTIAL INTER FUNDAMENTAL PARTICLE MOLECULAR CHEMICAL INTERACTIONS CHAINS AND CYCLES.

THIS IS PERIODICAL CYCLES OF REGENERATIONS OF ELECTRONS NUCLEONS AND ATOMS PARTICLE COMPOUNDS, WHICH IT ALTERNATE DURING EACH GIVEN ONE HALF LIFE OF ELECTRONS, NUCLEONS AND ATOM WITH DEGENERATIVE CYCLES, DEGENERATIONS THAT TAKE PLACE IN EVERY ONE HAF LIFE OF ATOMS, THERE AFTER THIS PHENOMENON CONTINUES UNDER DEGENERATIVE A. S. I. F.P. Mol. C.I.C. AUTONOMOUSLY IN CYCLES, THROUGH THESE CHEMICAL INTERACTIONS ENTIRE ELECTRONS NUCLEONS AND ATOMS PARTICLE COMPOUDS STRUCTURES REMOVED, AND IN THEIR PLACE AOTHER NEW STRUCTURES COPIES OF THE PRE-EXISTING ORIGINALS RE-CONSTRUCTED.

THE REGENERATION CYCLES OF TURN OVER PHENOMENON OF ELECTRONS, NUCLEONS AND ATOMS ARE EQUAL TO EVOLUTION ERA RE-GENESIS OF THE ELECTRONS, NUCLEONS AND ATOMS DURING PAST MOLECULAR EVOLUTION ERA, THAT IT TOOK PLACE IN THE MATTERS OF MILLIONS YEARS UNTIL THE PRESENTLY EXISTING ELECTRONS, NUCLEONS, AND ATOMS CONSTRUCTED AS STABLE STRUCTURES WE HAVE TODAY.

THIS TURN-OVER PHENOMENON OCCUR SIMILARLY IN ANOTHER FIELDS AS WELL, AT MICRO-UNITS CONSTRUCTIONS, ALSO IN MACRO-UNITS FIELDS OF CONSTRUCTIONS AS WELL, FOR EXAMPLE THE WHOLE ANIMAL BODY AND CELL POPULATION PERIODICALLY WITHIN SPECIFIC GIVEN HALF LIVES OF TURN OVER PROCESS, THE ENTIRE OLD STRUCTURES REMOVED, AND IT ALTERNATE AND REPLACED WITH RE-CONSTRUCTIONS AND REGENESIS OF CELLS AND BODY STRUCTURES WITH EXACT SAME COPIES AT REGENERATION CYCLES PERIDICALLY WITHIN SPECIFIC TIME INTERVALS.

MOLECULAR EVOLUTION

CYCLES OF AUTONOMOUS SEQUENTIAL CHEMICAL INTERACTIONS, AND CREATION OF LIFE

PHASE –ONE OF EVOLUTION

A. S. I. F.P. Mol. C.I.C. AND GENESIS OF NANO-UNITS:

THE ELECTRON-GENESIS, NUCELON-GENESIS, ATOM-GENESIS, ETC. (THE NANO-UNIT-GENESIS):

THE FUNDAMENTAL PARTICES ARE THE UNIT OF MATTER, AFTER INCEPTION OF PLANET EARTH, AT FIRST STEP THE AUTONOMOUS SEQUENTIAL INTER FUNDAMENTAL PARTICLE MOLECULAR CHEMICAL INTERACTION CYCLES AND CHAINS (A. S. I. F.P. Mol. C.I.C.) TOOK PLACE BETWEEN THE EARTH'S INDIGENOUS DIFFERENT FUNDAMENTAL PARTICLES, AND CONSTRUCTED DIFFERENT ELECTRONS, NUCLEONS. POSITRONS, QUARKS, NANO-UNITS AND ATOMS, ETC. THERE WAS NOT NANO UNITS CONSTRUCTIONS BEFORE THE FUNDAMENTAL PARTICLES AT EARTH.

THE PRESENTLY FEW DISCOVERED NANO-UNIT CLASSES SUCH AS ELECTRONS, NUCLEONS, ATOMS, POSITRONS, ETC. AT EARTH, ARE ONLY A FEW SMALL NUMBERS FROM NANO-UNITS CLASSES WHICH HAS BEEN DISCOVERED AT EARTH PRESENTLY, THERE ARE LARGE NUMBERS OF ANOTHER NANO-UNIT CLASSES, WHICH NOT BEEN DISCOVERED, AND NEEDS DISCOVERY IN FUTURE BY OTHERS.

PHASE –TWO OF EVOLUTION

A. S. I. N.-U. Mol. C.I.C. AND GENESIS OF BIOMOLECULES:

AT SECOND STEP THE AUTONOMOUS SEQUENTIAL INTER NANO-UNIT BASE MOLECULAR CHEMICAL INTERACTIONS CYCLES AND CHAINS (A. S. I. N.-U. Mol. C.I.C.) TOOK PLACE BETWEEN THE NEWLY CREATED DIFFERENT ATOMS AND ANOTHER NANO-UNITS, MOLECULAR STRUCTURES, AND DIFFERENT SIZES OF LARGER ATOM CONSTRUCTIONS, BIOMOLECULES, ETC. CREATED DIFFERENT SIZES MOLECULAR STRUCTURES EXISTED IN GENERAL CHEMISTRY, ORGANIC CHEMISTRY AND BIO-CHEMISTRY TEXTS AND MANY MORE.

PHASE –THREE OF EVOLUTION

A. S. I. C.B. Mol. C.I.C. AND GENESIS DIFFERENT CELLS, TISSUES, CREATURES, ETC:

THERE AFTER THE GENESIS OF DIFFERENT BIOMOLECULES, AND CELL, THE AUTONOMOUS SEQUENTIAL INTERNAL CELL AND BIOMOLECULAR COMBINATIONS AND CHEMICAL INTERACTIONS CYCLES AND CHAINS, TOOK PLACE BETWEEN THE PRODUCED DIFFERENT BIOMOLECULES, CELLS, TISSUES, ETC. AND CONSTRUCTED MORE DIFFERENT OTHER SPECIES CELLS, TISSUES, ORGANS, SYSTEMS, ETC. AND PRODUCED INNORMOUS DIFFERENT SPECIES OF LIVING THINGS FROM PLANTS TO ANIMALS, WHICH THE LAST ONE OF THOSE MOST ADVANCED ANIMALS BELONG TO THE HUMAN SPECIES, THIS SPECIES TODAY OVER GROWN ALL OVER THE EARTH, FOR THEIR DESCRIPTIONS SEE BIOLOGICAL TEXTS BOOKS.

ALTERNATING STABLE STEM CELL'S MUTATIONS WITH MEDIA CHEMICAL FORMULARY CHANGES

CONSTRUCTION OF ANIMALS AND PLANTS IN-UTERO FETUC (FETUS –GENESIS PHENOMENON) UNDER

ALTERNATING MUTATION SEQUENCES (AMS), WHEN ONE STABLE MUTATION ALTERNATE WITH ANOTHER

STABLE MEDIAL CHIMICAL FORMULARY CHANGES, IN UTERO AUTONOMOUSLY SEQUENTIALLY WITH EXPONENTIAL SPEEDS OF MULTIPLICATIONS AND PRODUCTIONS BETWEEN THE STEM CELLS, WHEN RAPIDLY ONE STABLE MUTATION STEM CELLS MUTATE INTO ANOTHER STABLE STEM CELL WITH EXPONENTIAL SPEED OF INTERACTIONS UNDER A. S. I. N.-U. –C.I.C.

ALTERNATING SEQUENTIAL STEM CELL MUTATIONS: THE MUTATIONS SEQUENTIALLY ALTERNATING WITH STEM CELLS MEDIA CHEMICAL FORMULARY CHANGES, IS CAUSE OF NEW TISSUE-GENESIS (MUTATIONS), IT IS CAUSE OF NEO- ORGAN - GENESIS, NEO-SYSTEM GENESIS, AND NEW FETUS CONSTRUCTION IN ANIMALS AND PLANT EMBRYONIC STATES. THIS IS PHENOMENON OF COMPLETE FETUS –GENESIS IN EMBRYO, THROUGH AUTONOMOUS SEQUENTIAL ALTERNATING MUTATION SEQUENCES, WHICH THEIR ALTERNATE WITH M.C.F.C., AND PRODUCE DIFFERENT TISSUES AND CELLS (CAUSE CELL DIFFERENTIATIONS), THIS IS THE CAUSE HOW ANIMALS AND PLANTS BODY CONSTRUCT A WHOLE CREATURE.

IN PLANTS THE MEDIA C.F.C. (CHANGING ONE COMBINING LIGHT PARTICLES TO OTHER) UNDER EVOLUTION PATH, CAUSE PLANTS STEM CELLS MUTATE AND CHANGE TO OTHER KIND STEM CELLS. THE MEDIA CHEMICAL FORMULARY CHANGES (M.C.F.C. = Y.F.P CHANGE), CAUSE STEM CELLS TRANSFORMS (MUTATIONS) THROUGH NEO PARTICLE COMPOUND GENESIS PHENOMENON FROM ONE TO OTHER, UNDER EVOLUTION ORDERS.

OPPOSITE OF ABOVE INTERACTIONS ALSO IS TRUE, THE CREATED NEW STEM CELLS MUTATION PRODUCE M. C. F. C., ONE FOLLOW OTHER IN CYCLES.

---

1 - STEM CELL'S MUTATIONS CAUSE, CHEMICAL FORMULARY CHANGES OF STEM CELL'S MEDIA (OR: M. C. F. C.) BECAUSE OF CELL'S METABOLISM.

2- M. C. F. C. (CHANGING MEDIA LIGHT PARTICLE IN AIR) CAUSE STEM CELL MUTATIONS (MUTATION OF EARLY STAGE FAST GROWING STEM CELLS). IN EMBRYO, THE GREEN COLOR LIGHT PARTICLE WAS ONLY COMBINING LIGHT PARTICLES, FOLLOWING OPENING CUSPIDS ALL OTHER LIGHT PARTICLES ALSO DIRECTLY INTER INTO SUBSYSTEM UNITS LAB.S AND COMBINE WITH PARTICLE COMPOUND, CHANGE THE CELL'S COLOR.

3- AT OPENING CUSPIDS, THE FLOWERS EXPOSE TO DIRECT OTHER COLOR LIGHT PARTICLES, DIFFERENT LIGHT PARTICLES INTER AND COMBINE WITH FLOWERS INTER ELECTRON NUCLEON PARTICLE COMPOUNDS, PRODUCE PARTICLE COMPOUND CHANGE IN FLOWER TISSUES (MUTATION GENESIS).

THIS IS MUTATION OF STEM CELLS FROM ONE LIGHT PARTICLE COMPOUND, TO ANOTHER PARTICLE COMPOUND. THIS PHENPMENON IS NEO- STEM CELL GENESIS UNDER A. S. I. F.P. Mol. C.I.C., ALSO BECAUSE OF PHENOMENON

OF ALTERNATING CELL'S MEDIA CHEMICAL FORMULARY CHANGES WHICH ALTERNATE WITH SEQUENTIAL STEM CELL NEO-MUTATIONS. THIS ALTERNATIONS ARE MAIN CAUSES OF CELL DIFFERENTIATIONS IN EMBRYO AND CONSTRUCTIONS OF WHOLE PLANT TISSUES AT EMBRYONIC PLANT STATES.

4- IN ANIMALS AND HUMAN EMBRYO THE SEQUENCES OF STEM CELLS MUTATIONS ALTERNATIONS WITH MEDIA CHEMICAL FORMULARY CHANGES ALTERNATION IS THE MOST IMPORTANT STEM CELLS DIFFERENTIATION FACTORS, AND MAIN CAUSE FOR DIFFERENT TISSUE-GENESIS, DIFFERENT ORGAN-GENESIS, DIFFERENT SYSTEM GENESIS AND DIFFERENT ANIMAL TOTAL BODY GENESIS DURING THE IN- UTERO FETUS DEVELOPMENTS AND GROW TO COMPLETE WHOLE WELL CONSTRUCTED NEW ANIMAL BABY READY TO BE DELIVERED TO OUTSIDE, ABOVE PHENOMENON IS ANIMAL GENESIS PHENOMENON INSIDE UTERUS ABOVE IS ONE EXAMPLE FROM CELL-BASE AND BIOMOLECULAR BASE MEDIA – C.F.C. WHEN IT IS ALTERNATING WITH STEM CELLS ALTERNATION PHENOMENONS.

PARTICLE BIOCHEMISTRY

STEM CELLS DIFFERENTIATIONS, OR STEM CELL'S ALTERNATING STABLE MUTATION SEQUENCES

AND PLANTS GENESIS (IN EARLY STAGE PLANT GROWTH PERIODS)

IN EARLY STAGE PLANT ATOM-CELL MUTATIONS, ONLY ARE UNDER A.S. I. F.P. Mol. C.I.C., AND MEDIA CHEMICAL FURMULARY CHANGES IS CHANGING INTERNAL ELECTRON NUCLEON PARTICLE COMPOUNDS, THESE CHANGES SECONDARILY PRODUCE ATOMS, CELL'S AND TISSUES CHANGES TO OTHER COLORS STRUCTURES (MUTATIONS), SUCH AS CHANGING COLORLESS CELLS TO GREEN CELLS, OR CHANGING GREEN CELLS TO DIFFERENT OTHER COLOR CELLS, THROUGH USE OF M.C.F.C., AS WELL THROUGH A.S. I. F.P. Mol. C.I.C.,

GREEN LIGHT PARTICLES COMBINATIONS UNDER A.S. I. F.P. Mol. C.I.C., AT FIRST SEQUENCE THEY PRODUCE GREEN COLOR MUTANT ATOMS, CELLS, TISSUES, AND PLANTS, IN ORDER TO CHANGE FIRST SEQUENCE MUTANTS, INTO OTHER COLORS LIGHT PARTICLES MUTANT STEM CELL'S PRODUCTIONS, IN PLANTS UNDER A. S. I. F.P. Mol. C.I.C. COMBINED WITH AIRBORN OTHER COLOR LIGHT PARTICLES, THEIR INTERNAL CELL'S ATOMS PARTICLE- COMPOUND CONSTRUCTIONS EXPLANIED IN THIS BOOK. BUT CELL BASE AND BIOMOLECULAR BASE MUTATIONS IN ANIMAL EMBRYO IS SUBJECT OF OTHER NEXT VOLUMES.

ATOM MUTATIONS, CELL MUTATIONS, AND PLANT MUTATIONS

THE AIRBORN, Y. g. – F.P. INTER INTO P.A.S. CURRENTS, AND FROM THERE THROUGH ELECTRON NUCLEONS P.C.S., INTER INTO INSIDE CLOSED BUDS, DIRECTLY INTER INTO INSIDE PETALS, OVARIES, STAMENS, INTERNAL ELECTRONS –NUCLEONS CHEMICAL LAB.S, AND COMBINE WITH INTERNAL ELECTRONS NUCLEONS PARTICLE COMPOUNDS, THROUGH REGENERATIVE A. S. I. F.P. Mol. C.I.C., PRODUCE THE GREEN COLOR LIGHT PARTICLE COMPOUNDS OF ENTIRE PETALS, OVARIES, STAMENS, OF EARLY STAGE FLOWERS BUDS, WHEN THE FLOWER CUSPIDS ARE THIGHTLY CLOSED, THROUGH REGENERATIVE A. S. I. F. P. Mol. C.I.C.THE CONSTRUCTED GREEN COLOR LIGHT PARTICLE COMPOUNDS CONSTRUCT MOST PARTICLE MOLECULAR CONSTRUCTIONS OF THE PETALS, OVARIES, STAMENS, ALL WITH GREEN COMPOUNDS, WHEN STILL THE CUSPIDS ARE CLOSED, THE ENTIRE CLOSED BUDS IN MOST SPECIES, AT EARLY GROWTH STAGES, CHANGE INTO GREEN COLOR CELLS AND TISSUES AND PETALS, CUSPIDS, OVARIES AND STAMENS, THIS IS PRIMARY SEQUENTIAL ATOM-CELL-TISSUE MUTATIONS IN PLANTS.

THE GREEN LIGHT FUNDAMENTAL PARTICLES, THROUGH THE SAME PATTERNS IN EARLY STAGE PLANT GROWTH, INTER INTO INTERNAL ELECTRONS, NUCLEONS OF FRUITS, NIDUS, ETC. AND COMBINE WITH FRUITS, PITS, INTERNAL ELECTRONS NUCLEONS PARTICLE COMPOUNDS, AND PRODUCE GREEN COLOR LIGHT PARTICLE _ COMPOUNDS, AND CONSTRUCT GREEN COLOR ATOMS, CELLS, BECAUSE THE CONSTRUCTED ELECTRONS AND NUCLEONS COLORS

ARE GREEN THROUGH THE GREEN COLOR LIGHT PARTICLE COMPOUNDS, WHICH ALL OF THOSE STRUCTURES ARE THE SAME COLORS OF COMBINING GREEN LIGHTS PARTICLES.

THIS IS THE REASON, ALL EARLY STAGE, PETALS, OVARIES, STAMENS, FRUITS, LEAVES, STEMS, ETC. ALL ARE GREEN COLOR, BECAUSE THEIR CONSTRUCTING PARTICLE COMPOUNDS ARE GREEN –LIGHT PARTICLE COMPOUNDS, WHICH THE CELLS AND TISSUES ADDITIONALLY TINT GREEN BECAUSE OF GREEN LIGHT PARTICLE COMPOUND CONSTRUCTIONS ALL TISSUES ARE GREEN COLORS.

PHENOMENON OF AUTONOMOUS SEQUENTIAL MUTATIONS, ALTERNATING SECODARILY TO MEDIA CHEMICAL FORMULARY CHANGES, AND VICE VERSA

TAKING PLACE IN LIVING THINGS EMBRYO, EQUAL TO MOLECULAR MUTATION

GENESIS OF LIVING THINGS

GENESIS OF THE LIVING THINGS IN EMBRYO IS EQUAL TO THE EVOLUTIO ORDERS, ANIMAL-GENESIS AND PLANT-GENESIS OCCUR BECAUSE OF TWO MAIN AUTONOMOUS SEQUENTIAL CHEMICAL FORMULARY CHANGES, WHICH ONE ALTERNATE THE OTHER, ONE CAUSE OTHER:

1 - MEDIA CHEMICAL FORMULARY CHANGES (M. C. F. C.) CAUSE, THE STEM CELLS MUTATIONS (S. C. M.).

2 - THE STEM CELLS MUTATIONS (S. C. M.) CAUSE, THE MEDIA CHEMICAL FORMULARY CHANGE (M. C. F. C.).

3 – THE M. C. F. C. ALTERNATE WITH EMBRYONIC CELL MUTATIONS. DURING EMBRYONIC INTERNAL UTERUS GROWTH, IS THE CAUSE FOR CELL DIFFERENTIATIONS IN ANIMALS,

EACH M. C. F. C. CREATE ANOTHER MUTATION, THE PRODUCED MUTANT CELL'S METABULISM SECONDARILY PRODUCE M. C. F. C. AGAIN IN CYCLE,

CLOSED CYCLES CHAINS OF MUTATIONS ALTERNATING WITH M. C. F. C. ONE AFTER THE OTHER DURING EMBRYONIC PHASES OF FETUS, CAUSE CREATIONS OF DIFFERENT TISSUES , ORGANS, BODY SYSTEMS, IN FETUS AN AFTER OTHER, AT THE END CONSTRUCT A FULL SIZE BABY WELL DEVELOPED AT THE BIRTH, THIS PHENOMENON IS ANIMAL –GENESIS AND PLANT GENESIS PHENOMENON IN EMBRYO, DEPENDING IT IS HAPPENING IN EARLY STAGE PLANTS STEM CELLS GROWTH, OR IT IS TAKING PLACE INSIDE ANIMAL EARLY STAGE STEM CELLS GROWTH, INSIDE THE UTERUS.

THIS IS THE PHENOMENON OF: MUTATIONS ALTERNATING WITH, M. C.F. C. INSIDE UTERUS, AND THIS IS THE CAUSE OF FETUS –GENESIS IN ANIMALS

FUNDAMENTAL PARTICLE'S BIOLOGICAL SCIENCES IN PLANTS

FUNDAMENTAL PARTICLE'S BIOLOGICAL CHEMISTRY SCIENCES, IN PLANTS

BIOCHEMISTRY OF FUNDAMENTAL PARTICLES IN PLANTS

COMBINATIONS OF EXOGENOUS LIGHT FUNDAMENTAL PARTICLES, WITH PLANTS INDOGENOUS INTER ELECTRON NUCLEON PARTICLE COMPOUNDS: AND NEO- PARTICLE COMPOUND GENESIS INSIDE PLANT'S ELECTRONS AND NUCLEONS:

AIRBORN EXOGENOUS SUN-LIGHT FUNDAMENTAL PARTICLES (EX. - Y.- F.P.), SHINE DIRECT ON PLANTS ELECTRONS – NUCLEONS, THE EX. –Y. –F.P. INTER INTO PLANT'S INTERNAL ELECTRON- NUCLEON SUBSYSTEM-UNITS CHEMICAL LAB.S, THE EXOGENOUS LIGHT PARTICLES COMBINE WITH INOGENOUS INTER ELECTRONS NUCLEONS PARTICLE COMPOUND'S (F.P. – COMP.) AND PRODUCE LIGHT FUNDAMENTAL PARTICLE COMPOUNDS (Y. - F.P. _ COMP.), THESE PRODUCED LIGHT PARTICLE COMPOUNDS CONSTRUCT PLANTS INTERNAL ELECTRON - NUCLEON PARTICLE COMPOUND CONSTRUCTIONS THROUGH A. S. I. F.P. Mol. C.I.C.

THIS PHENOMENON IS NEO- PARTICLE COMPOUND GENESIS, IT IS NEO-ELECTRON, NEO– NUCLEON –GENESIS PHENOMENON AS WELL.

DEPENDING WHAT COLOR EXOGENOUS LIGHT PARTICLES COMBINE WITH PLANTS INTERNAL ELECTRON NUCLEON PARTICLE COMPOUNDS, UNDER A. S. I. F.P. Mol. C.I.C., THE PRODUCED INTERNAL ELECTRON-NUCLEON LIGHT PARTICLE – COMPOUNDS COLORS WILL BE EXACTLY THE SAME COLOR AS THE COMBINING EXOGENOUS LIGHT PARTICLES COLORS, IF COMBINING EXOGENOUS LIGHT PARTICLE IS GREEN, THE PARTICLE COMPOUNDS COLORS WILL TINT GREEN AS WELL, THE RED COLOR OR YELLOW COLOR PETALS COMBINING LIGHT FUNDAMENTAL PARTICLES COLORS, RESPECTEDLY ARE RED – LIGHT PARTICLES AND YELLOW LIGHT PARTICLES COLORS.

THIS PHENOMENON IS THE CAUSE FOR PLANTS ELECTRONS, NEUTRONS, CELLS, ATOMS, COLOR CHANGES AFTER THE COMBINATIONS WITH EXOGENOUS FUNDAMENTAL PARTICLES, THE PLANT CELLS TINT WITH COMBINING LIGHT PARTICLES COLOR, THE PARTICLE COMPOUND COLORS CHANGES ALSO INTO COMBINING LIGHT PARTICLES COLORS.

THIS IS PHENOMENON OF COLOR – GENESIS IN PLANTS, IF COMBINING LIGHT PARTICLE IS GREEN COLOR THEREFORE PLANT ELECTRON-NUCLEONS PARTICLE- COMPOUNDS CELLS GETS GREEN COLOR AS WELL, (OR COLOR-GENESIS PHENOMENON IN PLANTS).

EARTH'S SENSIBLE LIGHT FUNDAMENTAL PARTICLES

MAJORITIES OF SUN- ORIGIN BIOFRIENDLY LIGHT FUNDAMENTAL PARTICLES CLASSES, ARE NON SENSIBLE AND NOT - DISCOVERED LIGHT PARTICLES. BIOFRIENDLY NON- SENSIBLE LIGHT PARTICLE CLASSES CONSTRUCT MOST OF PLANET EARTH'S INTER ELECTRONS - NUCLEONS PARTICLE COMPOUND CONSTRUCTIONS, ALSO INTERNAL MONO-ATOMS STRUCTURES, AND NANO-UNIT CONSTRUCTIONS, NANO-UNIT CLOUDS, AND ATOM – CLOUDS.

EARTH'S DISCOVERED SENSIBLE LIGHT FUNDAMENTAL PARTICLES CLASSES, PRESENTLY ARE KNOWN AS LIGHT-WAVES, THE SENSIBLE LIGHT FUNDAMENTAL PARTICLES AT EARTH ARE: GREEN –COLOR LIGHT PARTICLES, RED – LIGHT PARTICLES, AND YELLOW COLOR LIGHT PARTICLE CLASSES, FROM OTHER SENSIBLEE PARTICLES ARE, BLUE COLOR FUNDAMENTAL PARTICLES, OR VIOLET COLOR LIGHT PARTICLE CLASSES, WHICH ALL OF THESE PARTICLES ALSO CONSTRUCT PLANET EARTHS ATOM'S LECTRON – NUCLEON PARTICLE COMPOUND CONSTRUCTIONS.

PARTICLE BIOCHEMISTRY

THE COLOR OF PLANT'S PARTICLE –COMPOUNDS, WHEN PLANT'S ELECTRON-NUCLEON COMPOUNDS NOT COMBINED WITH LIGHT-PARTICLES.

PLANT'S COLOR IS WHITE CLEAR WATER COLOR, WHEN PLANT HAS NO LIGHT PARTICLE COMBINATIONS

IN CASES WHEN PLANTS TISSUE'S, POSSESS NO CHEMICAL COMBINATIONS WITH AIRBORN LIGHT-FUNDAMENTAL PARTICLES, MOSTLY IN THESE CASES, THE CELL'S COLORS, THE ATOM'S COLOR, THE PLANT'S TISSUES COLOR ALL STAY COLORLESS, OR WATER- CLEAR WHITE COLOR, IN MOST TISSUES WHICH, THESE CELLS DO NOT POSSESS, OR MINIMALLY POSSESS LIGHT PARTICLES COMBINATIONS WITH INTER ELECTRON – NUCLEON PARTICLE COMPOUNDS, AND THERE IS ONLY MINIMAL OR, LESS LIGHT PARTICLE COMPOUNDS INSIDE THEIR PLANT'S ELECTRONS NUCLEONS.

THESE CASES MOSTLY OCCUR AT EARLY STAGES, WHEN IN THIGHTLY CLOSED THICK FLOWER BUDS, OR INSIDE THICK NIDUS, WHEN THE LIGHT FUNDAMENTAL PARTICLES ARE UNABLE TRANSIT INTO INSIDE UN-PENETRABLE CLOSED PLANT SPACES.

AT EARLY GROWTH STAGE OF THIGHTLY CLOSED THICK BUDS, WHICH DOES NOT ALLOW THE LIGHT PARTICLES INTER INTO THIGHTLY CLOSED THICK BUDS, AND DOES NOT ALLOW LIGHT PARTICLES CHEMICAL COMBINATIONS WITH INTER ELECTRONS NUCLEONS PARTICLE – COMPOUNDS TO TAKE PLACE, THIS PHENOMENON ALSO OCCUR INSIDE HARD SHELLS OF NUTS AT EARLY STAGE GROWTH, WHICH LIGHT PARTICLES IS UNABLE CROSS THICK SHELLS AND COMBINE WITH PIT CELL'S ELECTRONS NUCLEONS, THESE TISSUES STAY WHITE COLORLESS CLEAR BECAUSE TISSUES ELECTRONS NUCLEONS POSSESS NO COMBINATIONS WITH LIGHT PARTICLES.

PLANT'S ATOMS –GENESIS, PLANT'S CELL- NEO-GENESIS BY GREEN LIGHT PATICLES

PLANT'S ELECTRON – NUCLEON PARTICLE COMPOUND – GENESIS, WITH GREEN LIGHT PARTICLE COMBINATIONS:

IN EARLY SPRING, WHEN THE EARLY PLANT STEM CELLS MULTIPLY WITH EXPONENTIAL SPEEDS, AND IN MANY SPECIES THE EARLY FLOWER'S THICK BUDS ARE THIGHTLY CLOSED, THE LIGHT PARTICLES ARE NOT CAPABLE FOR DIRECT INTERANCES INTO INSIDE CLOSED CUSPIDS, PETALS, OVARIES, STAMEN STAY COLORLESS.

ONLY THE CLOSED BUDS CUSPIDS, UNDER DIRECT DOMINANT Y. g. – F.P. INTERY INTO INSIDE CUSPID'S INTER ELECTRONS NUCLEONS SUBSYSTEM UNITS CHEMICAL LAB.S, HAVE BEEN COMBINED WITH INTER CUSPIDS INTERNAL ELECTRON-NUCLEON PARTICLE COMPOUNDS, AND HAVE PRODUCED GREEN LIGHT PARTICLE COMPOUNDS, AND THE Y. g. – F.P. COMPOUNDS HAVE BEEN CONSTRUCTED, THE CUSPIND'S ELECTRONS AND NUCLEONS MOLECULAR CONSTRUCTIONS WITH GREEN COLOR LIGHT PARTICLE COMPOUNDS, THROUGH THE A. S. I. F.P. Mol. C.I.C.

IN FIRST PHASE THE EXOGENOUS AIRBORN LIGHT FUNDAMENTAL PARTICLES, EXCEPT THE DOMINENT GREEN LIGHT FUNDAMENTAL PARTICLES ARE NOT CAPABLE TO TRANS – CROSS THE CLOSED CUSPIDS THICK WALLS, AND THIS CUSPID CLOSURE IS THE BEST PROTECTION PROCEDURE FOR OVARIES, STAMEN, PETALS OF THE EARLY STAGE FLOWER'S STEM CELLS.

AT THIS STAGE ONLY DOMINANT GREEN LIGHT PARTICLES CURRENTS, ARE CAPABLE DIRECTLY TO TRANS-CROSS THICK CUSPIDS, THROUGH P.A.S. –PARTICLE CIRCULATION SYSTEMS INTO INSIDE BUD'S CLOSED CAVITY, AND COMBINE WITH INTERNAL ELECTRON-NUCLEON PARTICLE COMPOUNDS, AND PRODUCE GREEN LIGHT PARTICLES – COMPOUNDS CONSTRUCTIONS INSIDE ELECTRON'S- NUCLEON'S OF PETALS, OVARIES, STAMENS, WHEN THE BUDS, CUSPIDS ARE CLOSED. AT THIS STAGE, THE PETALS, STAMENS, OVARIES PARTICLE COMPOUND CONSTRUCTIONS MOSTLY ARE CONSTRUCTED WITH GREEN LIGHT PARTICLE COMPOUNDS, ENTIRE FLOWERS CUSPIDS, PETALS, OVARIES, AND STAMEN ALL ARE GREEN COLOR.

THIS IS PHENOMENON OF PARTICLE COMPOUND NEO-GENESIS, THE ELECTRON NEO- GENESIS, THE NUCLEON NEO -GENESIS PHENOMENON.

AT THIS STAGE THE OTHER LIGHT FUNDAMENTAL PARTICLES ARE RECESSIVE, THEY ARE NOT CAPABLE TRANS – CROSS THROUGH P.A.S. CURRENTS CIRCULATION SYSTEMS, INTO INSIDE CAVITY OF THIGHTLY CLOSED, THICK CLOSED CUSPIDS, (ONLY DOMINANT Y. g. –F.P. POSSESS THIS CAPABILITY).

PLANT GENESIS PHENOMENON

THE FUNDAMENTAL PARTICLE BIOCHEMISTRY

IN STEM CELLS RAPID GROWTHS PHASES IN EARLY SPRING, WHEN FLOWER BUDS OPENS, IMMEDIATELY PETALS, STAMEN, OVARY ARE UNDER DIRECT EXPOSURE OF ALL DIFFERENT COLOR LIGHT FUNDAMENTAL PARTICLES, AT THIS PHASE ALL LIGHT PARTICLES SUCH AS, RED, YELLOW, BLUE, VIOLET, COLOR LIGHT PARTICLES DIRECTLY AIRBORN INTER INTO INSIDE PETALS, OVARIES, STAMENS, ELECTRONS – NUCLEONS SUBSYSTEM-UNITS CHEMICAL LAB.S, AND COMBINE DIRECT WITH INTER ELECTRON –NUCLEON PARTICLE COMPOUNDS, AND PRODUCE LIGHT PARTICLE COMPOUNDS WHICH THOSE NEWLY PRODUCED LIGHT PARTICLE COMPOUNDS COLORS, ARE EXACTLY THE SAME COLOR OF COMBINING INCIDENT EXOGENOUS LIGHT PARTICLE COLOR, THAT COMBINED WITH INTERNAL ELECTRON NUCLEON PARTICLE COMPOUNDS.

PRODUCED FLOWER ATOMS, CELLS, TISSUES ARE IN EXACT COLORS OF INCIDENT EXOGENOUS COMBINING LIGHT FUNDAMENTAL PARTICLE'S COLORS. THESE PRODUCED NEW PARTICLE COMPOUNDS, WHICH HAS DIFFERENT COLORS FLOWER CONSTRUCTIONS, AND PLANT'S INTERNAL ELECTRON- NUCLEON PARTICLE COMPOUND CONSTRUCTIONS ARE DIFFERENT COLORS, THIS IS LIGHT PARTICLE –COMPOUND GENESIS PHENOMENONS, NEO - ELECTRON- NUCLEON GENESIS PHENOMENONS, ETC.

AS EXPLAINED BEFORE AT FIRST SEQUENCES A.S. I. F.P. Mol. C.I.C. IN EARLY STAGES PLANTS GROWTH THERE IS GREEN LIGHT PARTICLES COMBINATIONS DOMINANCE WITH MOST OF FLOWERS FRUITS AND PLANTS ELECTRONS AND NUCLEONS, AT EARLY STAGES STEM CELL RAPID GROWTHS PLANTS STAGES NOT ONLY ATOM BASE AUTONOMOUS SEQUENTIAL INTERACTIONS AND STEM CELL BASE SEQUENTIAL CHEMICAL INTERACTIONS TAKE PLACE AT EXPONENTIAL SPEED RATES, AT THE SAME TIME SUN LIGHT FUNDAMENTAL BASE A. S. I. F.P. Mol. C.I.C. ALSO ALL RUNING WITH EXPONENTIAL SPEEDS OF CHEMICAL INTERACTIONS, AND CONSTRUCTING DIFFERENT PLANTS TISSUES INTERNAL ELECTRON NUCLEON CONSTRUCTIONS, THROUGH EXPONENTIAL SPEEDS IN ALL CELL- ATOM- PARTICLE BASE CHEMICAL SEQUENCES WITH COOPERATIONS AND COORDINATIONS OF EACH OTHER, IN SHORT TIME EMBRYONIC EARLY SPRING STEM CELL STAGE, AT EARLY SPRING EMBRYONIC STEM CELL EXPONENTIAL GROWTH PERIOD GREEN LIGHT PARTICLE COMBINE WITH ALL PLANTS TISSUES AT OVER 85 % OF EXISTING PLANTS SPECIES.

UNDER SECOND SEQUENCES OF A. S. I. F.P. Mol. C.I.C. IN LATER STAGES, WHEN FLOWERS OPENS AND THERE IS DIRECT EXPOSURES OF FLOWER ATOMS TO ALL LIGHT PARTICLES, AT THIS STAGE THE PRE-CONSTRUCTED GREEN COLOR LIGHT PARTICLE COMPOUNDS UNDER DEGENERATIVE A. S. I. F.P. Mol. C.I.C.AT INSIDE FLOWERS FRUITS PLANTS PETALS-OVARIES-STAMEN, AS WELL AS AT FRUITS ETC. ALL BREAK DOWN INTO SMALLER CONSTRUCTING MOLECULAR STRUCTURES, SUCH AS FREE GREEN LIGHT PARTICLES, WHICH GREEN LIGHT PARTICLE ALSO RELEASE AS FREE AIRBORN PARTICLES SCAPE TO AIR AND DISAPPEAR.

IN SECOND SEQUENCES THGERAFTER, THE PLANTS FLOWERS, FRUITS, PARTICLE COMPOUNDS, ETC. THEY ARE NOT GREEN COLOR ANY MORE, AND THEY HAVE BEEN COMBINED WITH OTHER COLOR LIGHT PARTICLES, AND HAVE CONSTRUCTED THE ELECTRONS NUCLEONS CONSTRUCTIONS WITH ANOTHER COLOR LIGHT PARTICLE COMPOUDS, AND FLOWERS, PETALS, STAMENS, FRUITS, ETC. HAVE BEEN CHANGEED COLORS TO THE ANOTHER ONES, SUCH AS RED, YELLOW, VIOLET, BLUE COLORS, DEPENDING WHICH COLOR LIGHT PARTICLE COMBINED WITH WHICH PART ATOMS OF THE PLANTS LEAVES, FLOWERS, FRUITS, STEMS, ETC. HAVE THE CHEMICAL COMBINATIONS, UNDER THE A.S. I. F.P. Mol. C.I.C.

WHEN CUSPIDS AND FLOWERS OPEN IN LATER STAGES, THERE AFTER OTHER COLOR LIGHT PARTICLE SUCH AS RED, YELLOW, BLUE, VIOLET, ETC. THROUGH DIRECT SUN SHINE, INTER INTO FLOWERS FRUITS PLANTS INTER ELECTRONS NUCLEON SUBSYSTEM – UNITS CHEMICAL LABS. AND UNDER A. S. I. F.P. Mol. C.I.C. COMBINE WITH INTER ELECTRONS NUCLEONS PARTICLE COMPOUNDS AND PRODUCE DIFFERENT COLORS LIGHT PARTICLE COMPOUNDS

OF RED, YELLOW, BLUE, VIOLET, OR OTHER COLOR PARTICLE COMPOUNDS, PARTICLE COMPOUNDS COLORS ARE THE SAME AS COMBINING LIGHT PARTICLES, ALSO THE ATOMS, CELLS COLORS ALSO ARE THE SAME AS THEIR CONSTRUCTING PARTICLE COMPOUNDS.

DEPENDING WHICH COLOR LIGHT PARTICLE COMBINING TO FRUITS, FLOWERS, PLANTS PARTICLE COMPOUNDS, THEY PRODUCE THAT COLOR LIGHT PARTICLE COMPOUNDS INSIDE PLANTS FLOWERS, FRUITS, ELECTRONS, NUCLEONS, ATOMS, THE COLOR OF PLANT CHANGE INTO COMBINING LIGHT PARTICLES COLORS.

THIS PHENOMENON IS PLANT- GENESIS PHENOMENON, AND PHENOMENON OF TRANSFORMATIONS OF ONE COLOR PLANT TISSUES TO THE OTHER COLORS PLANTS, THESE PHENOMENONS ALL ARE STABLE MUTATIONS WHICH ARE TAKING PLACE UNDER PREVIOUS EVOLUTION ORDERS OF STABLE MUTATIONS, SEQUENCES, ONE AFTER OTHER.

THE INTELLIGENT PLANTS, NON INTELLIGENT PLANTS

PLANTS CENTRAL INTELLIGENCE SYSTEMS CENTERS (C. I. S.)

PLANTS PERIPHERAL INTELLIGENCE SYSTEM CENTERS (P. I. S.)

THERE ARE TOO MANY PLANT SPECIES, THAT DO NOT HAVE PLANT INTELLIGENCE SYSTEM CENTERS, THERE ARE SOME PLANTS ONLY POSSESSES PARTIAL AND INCOMPLETE CENTRAL INTELLIGENCE SYSTEM CENTERS, THERE ARE ALSO LARGE NUMBERS OF PLANT SPECIES THAT POSSESSES BOTH THE CENTRAL INTELLIGENCE SYSTEM CENTERS (C. I. S.), AND PERIPHERAL INTELLIGENCE SYSTEM CENTERS (P. I. S.) BOTH.

THERE ARE TWO MAIN INTELLIGENCE SYSTEM CENTERS IN PLANTS:

1 – THE PLANT'S PRIMORDIAL CENTRAL INTELLIGENCE SYSTEM CENTERS (C. I. S.) OR PLANT'S MAIN INTELLECTUAL COMMAND CENTERS SYSTEMS.

2 – THE PLANT'S PERIPHERAL INTELLIGENCE SYSTEM CENTERS (P. I. S.):

THE PERIPHERAL INTELLIGENCE SYSTEM CENTERS ALSO DEVIDES INTO TWO DIFFERENT INTELLIGENCE SYSTEMS:

1 – THE PLANTS SENSARY INTELLIGENCE SYSTEM CENTERS (S. I. S.).

2 – THE PLANTS MOTOR INTELLIGNCE SYSTEM CENTERS (M. I. S.).

THE BIOFRIENDLY VISIBLE OR NON-VISIBLE LIGHT FUNDAMENTAL PARTICLES HAVE VITAL IMPORTANCE FOR PLANTS, THE LIGHT FUNDAMENTAL PARTICLES DAILY SHINE ON FLOWERS, LEAVES, STEMS, FRUITS, DIRECT INTER INTO INTERNAL ELECTRON NUCLEON SUBSYSTEM –UNITS CHEMICAL LAB.S, COMBINE WITH INTER ELECTRON NUCLEON PARTICLE COMPOUNDS, CONSTRUCT PLANTS ELECTRONS AND NUCLEONS LIGHT PARTICLE COMPOUND CONSTRUCTIONS, UNDER THE REGENERATIVE A. S. I. F.P. Mol. C.I.C. (NANO-UNIT GENESIS PROCESS).

THE PLANT'S SENSARY SYSTEM CENTERS, CONTINUOUSLY MONITOR AND REPORT EXISTANCES OF ENVIRONMENTA SUN SHINE, OR EXISTING ENVIRONMENTAL HAZARDS, SUCH AS BURNING SUN IN MID SUMMERS WHICH MAY CAUSE DEHYDRATION AND SUN BURN, OR EXISTANCES OF FREEZING COLD IN LATE WINTER, OR IN EARLLY SPRING WHICH THE FLOWERS OPEN MAY DIE, THROUGH FROST BITES, THESE ENVIRONMENTAL INFORMATIONS THROUGH PARTICLE CLOUDS TRANSIT, TO THE HIGHER CENTRAL INTELLIGENCE SYSTENS CENTERS (C.I.S.) FOR THEIR DECISIONS, AND ISSUING THE PROPER TASK – ORDERS, FOR MOTOR INTELLIGENCE SYSTEM CENTERS EXECUTIONS.

FOR EXAMPLE, THE CIS IN DAILY SUN SHINE MAY ORDER THE PLANTS FLOWERS, LEAVE STEMS TURN OR TILT TOWARD THE INCOMING SUN LIGHT DIRECTIONS, OR IN THE CASE OF EXISTING FREEZING COLD, MAY ORDER THE CUSPIDS TO CLOSE AND PROTECT THE PETALS FROM FROST BITS, OR IN BURNING HOT TEMPERATURE THE CIS MAY ORDER THE FLOWERS TO TURN DOWN OR AWAY FROM THE INCOMING SUN LIGHT DIRECTIONS TO REDUCE DEHYDRATIONS AND BURNS, ETC.

THE PLANTS C. I. S. CENTERS, THROUGH THE PLANTS MOTOR INTELLIGENCE SYSTEM CENTERS, MAY C. I. S. ORDERS TO MOTOR INTELLIGENCE SYSTEM CENTER THAT, THE FLOWERS, THE LEAVES, AND STEMS MUST ROTATE AND TURN TOWARD THE DIRECTION OF INCOMING LIGHT FUNDAMENTAL PARTICLES DAILY WHEN THERE IS SUN SHINE IN ORDER TO CAPTURE THE MAXIMUM AMOUNT NEEDED LIGHT FUNDAMENTAL PARTICLE FROM AIR, AND DELIVE INTO ELECTRONS NUCLEONS CHEMICAL LAB.S, TO BE USED IN A.S.I.F.P. Mol. C.I.C. FOR RE-GENESIS OF NEEDED PLANTS NANO-UNIT STRUCTURES. THE MAIN FUNCTION OF SENSARY INTELLIGENCE SYSTEM CENTERS IS TO REPOR EXISTING SUN SHINE, OR EXISTING ENVIRONMENTAL HAZARDS SUCH AS FREEZING COLD, OR BURNING HOT SUMMER, ETC. TO THE HIGHER COMMAND INTELLIGENCE SYSTEM CENTERS.

THE ESSENTIAL FUNCTIONS OF PLANTS COMMAND SYSTEMS CENTERS IS TO OPERATE PROTECTIVE AND DEFENSIVE MEASURES, PROVIDE AND MAINTAIN THE CHEMICAL, PHYSICAL, AND BIOLOGICAL NEEDED TASKS, IN NORMAL LEVELS IN THE PLANTS. HELP PLANTS CAPTURE NEEDED LIGHT FUNDAMENTAL PARTICLES, ORDER AND ACHIEVE NORMAL PLANTS FUNCTIONS, AND A. S. I. F.P. Mol. C.I.C., ETC., THE PLANTS C.I.S. -CENTERS IN DAILY BASIS, ORDER THE PLANTS LEAVES, FLOWERS, STEMS, TO TURN TOWARD INCOMING LIGHT PARTICLE DIRECTIONS, TILT, OR ROTATE, AND TURN TOWARD LIGHT PARTICLE DIRECTIONS WHEN THERE IS SUNSHINE. IN ORDER TO COLLECT LIGH PARTICLES FROM INCOMING DIRECTIONS OF LIGHT ORIGIN PARTICLES, THROUGH THE FLOWERS AND LEAVES AND STEM ROTATIONS OR TILTS, UNDER THE CIS INTELLECTUAL COMMAND SYSTEMS CENTERS ORDERS THE LEAVES, FLOWERS, STEMS TURN TOWARD LIGHT PARTICLE DIRECTIONS, AND ATTRACT THE LIGHT PARTICLES.

IN DAILY BASIS, THE AIRBORN INCOMING LIGHT PARTICLES DIRECTLY INTER INTO INSIDE PLANTS INTER ELECTRON NUCLEON SUBSYSTEM –UNITS CHEMICAL LAB.S, AND UNDER A. S. I. F.P. Mol. C.I.C. COMBINE WITH INTER ELECTRON-NUCLEON PARTICLE COMPOUNDS AND CONSTRUCT LIGHT PARTICLE COMPOUNDS OF PLANTS ELECTRON NUCLEON CONSTRUCTIONS, (PHENOMENONS OF PARTICLE COMPOUND NEO- GENESIS AND PHENOMENON OF NEW ELECTRON- NUCLEON –GENESIS) ALL OF THESE NANO-TASKS, SENSED AND REPORTED BY S.I.S, AND NECESSARY ORDERS ISSUED THROUGH THE C.I.S., AND THE TASKS EXECUTED BY M.I.S CENTERS. ALL PLANTS DEFENSIVE AND PROTECTIVE MEASURES, DAILY PHYSICAL CHEMICAL BIOLOGICAL FUNCTIONS ALL SENSED BY S.I.S., ORDERS ISSUED BY C.I.S. CENTERS, AND ISSUED ORDERS EXECUTED BY M.I.S. CENTERS OF THE PLANTS. UNDER DIRECT C.I.S. CENTERS ORDERS THE FLOWERS, STEMS, LEAVES, TURN INTO DIRECTION OF SUN LIGHT, THE PLANTS DEFENSIVE AND PREVENTIVE ORDERS ALSO DONE BY CENTRAL INTELLIGENCE SYSTEM CENTERS AND ISSUED ORDERS CARRIED OUT BY M.I.S. CENTERS.

THE S.I.S. CENTERS FUNCTIONS ARE SENSING, AND REPORTING, EXISTING ENVIRONMENTAL CONDITIONS, AS WELL AS REPORTING EXISTANCES OF HARMFUL ENVIRONMENTAL CONDITIONS, SUCH AS DEHYDRATIONS, LACK OF IRRIGATIONS ALL GET REPORT TO C. I. S. CENTERS FOR THEIR DECISIONS AND ORDERS. AND CIS EXECUTE THE ORDERS THROUGH M.I.S. CENTERS.

THROUGH THE PROPER OPERATIONS OF S. I. S., C. I. S. AND M. I. S. CENTERS CLOSELY, THE CARNIVOROUS PLANTS CAN DO INSECT HUNTING THROUGH QUICK INTELLIGENCE SYSTEMS OPERATIONS. SENSARY SYSTEMS CAN PINPOINT EXACT INSECT LOCATIONS ON LEAVES, AND C.I.S. CAN ISSUE ORDER AT EXACT TIME, AND M. I. S. CAN EXECUTE THE HUNTING ORDERS ACCORDINGLY. AND CATCH FAST MOVING INSECT IN EXACT LOCATION AND EXACT MOMENTS, AND DELIVER THE HUNTED LIVE INSECTS FOR PLANTS NUTRITION AS ORDINARY DAILY FOOD.

THE PARTICLE CLOUD CURRENTS, AND PARTICLE CIRCULATION SYSTEMS (PCS) IN PLANTS

THE PRIMARY P. C. S. CONNECTS THE THREE SENSARY - CENTRAL - MOTOR INTELLIGENCE SYSTEM ELECTRONS-NUCLEONS TO EACH OTHER, THROUGH BI-DIRECTIONAL SEPARATE EFFERENT AND AFFERENT PARTICLE CIRCULATION CURRENTS, WHICH THESE TWO DIRECTION CURRENTS, ONE PARTICLE CLOUD CURRENTS RUNS, IN OPPOSITE DIRECTION OF THE OTHER PARTICLE CLOUD CURRENT.

ALSO, THE SECONDARY P. C. S. CONNECTS THE PLANTS: STEMS, FLOWERS, LEAVES, ROOTS, FRUITS TOTAL ELECTRON – NUCLEON POPULATIONS TO EACH OTHER. AND ALL OF THESE SECONDARY P. C. S. CURRENTS FINALLY CONNECTS TO MAIN PRIMARY PLANTS PARTICLE INTELLIGENCE SYSTEMS CENTERS ELECTRONS –NUCLEONS POPULATIONS P. C. S. CURRENTS FROM THE OTHER SIDE,

THROUGH THESE TWO DIFFERENT DIRECTION PRIMARY AND SECONDARY PARTICLE CLOUD CIRCULATIONS SYSTEMS CURRENTS THE ENTIRE PLANTS ELECTRONS NUCLEONS POPULATIONS GET CONNECTION TO EACH OTHER. ABOVE IS P.C.S CONSTRUCTIONS BRIEFLY. WHICH THROUGH THESE PCS THE PLANTS INTELLECTUAL CENTERS ARE ABLE TO MANAGE THE OPERATIONS OF THE ENTIRE PLANTS BODY ELECTRONS NUCLEONS FUNCTIONS.

CLASSIFICATIONS OF PLANT SPECIES UNDER PLANT'S PARTICLE INTELLIGENCE SYSTEMS:

UNDER THIS CLASSIFICATIONS THE DIFFERENT PLANTS SPECIES DIVIDE INTO THREE DIFFERENT CLASSES: EITHER PLANTS HAVE ALL THREE SENSARY- CENTRAL-MOTOR INTELLECTRUAL SYSTEMS CENTERS ALL. OR THE PLANTS SPECIES WHO DO NOT HAVE ANY KIND INTELLIGENCE SYSTEM CENTERS AT ALL. ALSO THERE ARE THIRD CLASS LARGE NUMBERS OF PLANTS SPECIES WHO POSSESSES INTERMEDIATE AND INCOMPLETE PLANT INTELLIGENCE SYSTEM CENTER, WHICH EXPLAING CHARACTORS OF THIS THIRD CLASS IS TOO LARGE, AND DO NOT FIT TO BE EXPLAINED HERE.

THE CLASSIFICATION OF PLANTS UNDER PARTICLE INTELLIGENCE SYSTEMS:

1 - THE PLANT SPECIES WHO POSSESS PC S AND PIS

2 - THE PLANTS SPECIES THAT DO NOT HAVE PCS AND PIS

3 - THE PLANTS SPECIES WHO POSSESS PARTIAL AND INCOMPLETE PCS, AND PIS,

THE PLANT SPECIES WHO POSSESS PCS AND PIS

PLANTS PRIMORDIAL INTELLIGENCE SYSTEMS CENTERS COMPOSED FROM THREE MAJOR DIFFERENT INTELLIGENCE SYSTEM CENTERS AS FOLLOWING,

1- PLANTS CENTRAL INTELLIGENCE SYSTEM CENTERS (C. I. S.), 2 - PLANTS SENSARY INTELLIGENCE SYSTEMS CENTERS (S. I. S.), 3 - PLANTS MOTOR INTELLIGENCE SYSTEMS CENTERS (M. I. S.),

THESE ADVANCED PLANT SPECIES POSSESS CAPABILITIES THROUGH SENSARY INTELLIGENCE SYSTEMS SENSE, CAPTURE AND RECEIVE LIGHT INFORMATION IMAGE PARTICLE CLOUDS FROM AIR IN REGARD TO OUTSIDE WORLD ENVIRONMENTAL CONDITIONS AND INSIDE CENTRAL INTELLIGENCE SYSTEM CENTERS ANALIZE THE SENSARY SYSTEMS REPORTED CONDITIONS AND DECIDE, ORDER, RESPOND ACCORDING TO CONDITIONS AND CARRY ORDERS TO PLANTS MOTOR INTELLECTUAL SYSTM FOR EXECUTION,

THE SENSARY INTELLIGENCE SYSTEMS ELECTRONS NUCLEONS THROUGH THEIR HIGH GRAVITON FORCES, ATTRACT AIRBORN LIGHT PARTICLES AND LIGHT FUNDAMENTAL PARTICLE INFORMATIONS- IMAGES PARTICLE- CLOUDS (Y. F.P. –I.I.- P.cl.) FROM AIR, AND TRANSIT INTO PLANTS CENTRAL INTELLIGENCE SYSTEM CENTERS ELECTRONS – NUCLEONS, THE INCOMING Y. F.P. –I.I.- P.cl. DIRECTLY INTER INTO INTER ELECTRON- NUCLEON SUBSYSTEM UNITS PLANTS CENTRAL INTELLIGENCE SYSTEMS AND COMBINE WITH INTER ELECTRON-NUCLEON PARTICLE COMPOUNDS UNDER A. S. I. F.P. Mol. C.I.C. AND PRODUCE Y. –F.P.- I.I. – P. cl. _COMPOUNDS, AND THESE PRODUCED PARTICLE CLOUD COMPOUNDS CONSTRUCT THE C. I. S. ELECTRONS NUCLEONS CONSTRUCTIONS, (NEO-ELECTRON-NUCLEON GENESIS).

THE DIFFERENT CIS PRE-EXISTING CONSTRUCTED ELECTRONS –NUCLEONS THROUGH THESE C. I. S. INFORMATIONS AND IMAGES DECIDE IN REGARD TO NEW INCOMING PARTICLE CLOUDS AND DIAGNOSE THE SITUATIONS AND ISSUE THE C.I.S. ORDERS TO M. I. S. ELECTRONS NUCLEONS TO BE EXECUTED ACCORDINGLY, THROUGH THESE COMPLEX A. S. I. F.P. Mol. C.I.C., PARTICLE CLOUD COMPARISIONS AND THE C.I.S. ELECTRONS-NUCLEONS DECISIONS OVER THE S.I.S. REPORTS.

OVER INFORMATIONS OF S. I. S. SENSORY PARTICLE CLOUDS REPORTS, THE C.I.S. DECIDES ITS ORDERS, AND ISSUE ORDERS TO M. I. S., THROUGH COMPARISON OF ENORMOUS PRE-EXISTING STORED Y. –F.P. – I.I. – P. cl. COMPOUNDS, WHICH ALL PREVIOUSLY HAVE BEEN STORED INSIDE THE DIFFERENT ELECTRONS NUCLEONS, SIMILAR AN ENCYCLOPEDIA OF INFORMATIONS, FOR COMPARISONS AND DECISION MAKING BY C. I. S. ELECRONS NUCLEONS.

THE PRODUCED ORDERS THROUGH OUTGOING F.P.- I.I. – P.cl. CURRENTS CIRCULATION SYSTEMS TRAVEL TO PLANTS MOTOR INTELLIGENCE SYSTEM CENTERS (M.I.S.) AND MOTOR INTELLIGENCE CENTERS EXECUTE C.I.S. - ORDERES ACCORDING ISSUED ORDERS THROUGH PLANTS LEAVES- FLOWERS- STEMS ALL ACT ACCORDING ISSUED ORDERS, ISSUED ORDERS MAKES FLOWERS LEAVES STEMS TURN ROTATE ACCORDING S. I. S. REPORTS AND C. I. S. ORDERS, AND M.I.S. SECODARY ORDERS OVER THE STEMS, LEAVES. FLOWERS, AND ROOTS THAT HOW THEY MUST ACT, ROTATE, TILT, OR NOT DO ANY THING.

PLANTS INTELLIGENCE SYSTEM'S FUNCTIONS

EXAMPLES FROM PHYSIOLOGICAL FUNCTIONS UNDER INTELLIGENCE SYSTEM CENTERS.

1 – IN LATE WINTER, OR EARLY SPRING, WHEN WARM WEATHER TRIGGER FLOWERING PLANTS, IN CASES OF SUDDEN FREEZING EPISODES, THE S. I. S. SENSE EVENTS, AND REPORT FREEZING COLD, POSSIBLE FROST BITE TO C. I. S., UNDER ANALYSIS AND DECISSIONS THE C. I. S. ORDERS TO THE M. I. S., TO CLOSE CUSPIDS, AND PROTECT PETALS, OVARIES, STAMEN, FROM FROST BITES, IN THE PLANT SPECIES WHO POSSESS INTELLIGENCE SYSTEMS CENTERS, AND M.I.S. CARRIES ORDERS OF C.I.S. ACCORDINGLY.

IN MANY PLANT SPECIES WHO DO NOT POSSESS INTELLIGENCE PLANT SYSTEMS CENTERS, THESE SPECIES FLOWERS STAY WIDELY OPEN IN FREEZING COLD, AND SUFFER FROST BITE AND DIE. BECAUSE THEY COULD NOT SENSE, BECAUSE THEY DO NOT POSSESS INTELLIGENCE SYSTEMS, TO SENSE AND CLOSE CUSPIDS, PROTECT PETALS OVARIES STAMEN FROM FROST BITES.

2 - IN NORMAL CLIMATE, WHEN THERE IS SUN SHINE, DIFFERENT LIGHT FUNDAMENTAL PARTICLES ARE PLENTY AROUND AIRBORN COMING FROM PLANET SUN, UNDER THESE CONDITIONS, AS PLANTS NEEDS ALL KIND LIGHT FUNDAMENTAL PARTICLES TO CONSTRUCT THEIR INTERNAL ELECTRON NUCLEON PARTICLE COMPOUND CONSTRUCTIONS, THE S.I.S. SENSE SITUATIONS AND REPORT STATUS TO UPPER INTELLIGENCE SYSTEM CENTERS, THE C.I.S. ISSUE ORDERS THAT ALL FLOWERS, LEAVES, FRUIT'S INTERNAL ELECTRON-NUCLEON PARTICLE COMPOUNDS CONSTRUCTIONS THROUGH CAPTURING NEEDED LIGHT PARTICLES FROM SURROUNDING AIR.

THE SENSARY SYSTEMS (S. I. S.) SENSE LIGHT PARTICLES, TEMPERATURES AND REPORTS PARTICLES DIRECTIONS, THROUGH Y. -F.P.- I.I. – P.cl. CURRENTS WHICH CARRIED TO HIGHER PLANTS CENTERAL INTELLIGENCE SYSTEMS CENTER (C. I. S.), IN RESPONSE TO S. I. S. FINDINGS, THE CENTERAL INTELLIGENCE SYSTEM CENTERS ISSUE PROPER ORDERS TO MOTOR INTELLIGENCE SYSTEMS (M. I. S.), TO EXECUTE ORDERS, TURNING FLOWERS AND LEAVES IN DIRECTION OF LIGHT PARTICLE TOWARD SUN, TO CAPTURE MAXIMUM NEEDED QUATITIES LIGHT PARTICLES FOR ELECTRON-NUCLEON –GENESIS WHICH ARE SURVIVAL FOR PLANT'S LIFE.

3 - IN EXTREME BURNING SUMMER TEMPERATURES, THE SENSARY INTELLIGENCE SYSTEMS SENSE DANGEROUS BURNING, DRYINGS, LETHAL, DEHYDRATING CONDITIONS, AND REPORT TO C. I. S. THE CENTRAL INTELLIGENCE SYSTEM CENTER ORDER TO MOTOR INTELLIGENCE SYSTEMS CENTERS TO TURN THE FLOWERS FACE DOWN, AND AWAY FROM EXPOSURE TO DIRECT HOT BURNING SHINING LIGHT PARITCLES, AND PREVENT AND AVOID FROM DIRECT BURNING LIGHT PARTICLES EXPOSURES.

IN FEW ANOTHER CASES OR SPECIES WHO POSSESS BETTER DEFENSIVE AND IMMUNITY SYSTEMS UNDER THESE CONDITIONS WHEN THE PLANTS RECEIVE PROPER HELPS SUCH AS IRRIGATIONS, ETC., FLOWERS, LEAVES START THEIR ORDINARY FUNCTIONS TURN TOWARD SUN TO COLLECT MORE LIGHT PARTICLES AND WHICH THEY NEEDS FOR DAILY USE,

IN SPECIES WHO DO NOT POSSESS THE PLANT INTELLIGENCE SYSTEM CENTERS, THESE SPECIES GET HARMED FROM HEAT STROKE, BURNS, DEHYDRATIONS, MUCH EASIER THAN THOSE WHO HAVE PRIMARY INTELLIGENCE SYSTEM CENTERS.

THE PLANT'S IMMUNITY SYSTEMS AND DEFENSIVE SYSTEMS

PLANT'S CHILDHOOD ERA, THEREAFTER PLANT'S AGING ERA EFFECTS, AND LIVING IN GERIATRIC AGES, AND GRADUAL SLOWING DOWN OF PLANTS INTELLIGENCE SYSTEMS, DEFENSE SYSTEMS, IMMUNITY SYSTEMS:

THE YOUNG HEALTHY FLOWERS AND PLANTS POSSESS MUCH BETTER DEFENSE SYSTEMS AND IMMUNITY OPERATIONS AND THEIR SENSARY SYSTEMS CENTRAL INTELLIGENNCE SYSTEMS AND MOTOR SYSTEMS ALSO FUNCTION SHARPER AND MUCH BETTER THAN OLDER OR SICK LOOKING PLANTS,

THE ABOVE PHENOMENON IN SOME SPECIES, ALSO ARE BETTER THAN THE OTHER SPECIES WHO DOT POSSESS OR NOT DEVELOPED THESE SYSTEMS. ADDITIONALLY, IN OLDER AGE PLANTS, EVEN THE CENTRAL INTELLIGENCE SYSTEMS OF PLANTS SLOWS DOWN, AND FUNCTIONS CAN NOT BE COMPARED TO YOUNG AGE PLANTS AT ALL.

IN EARLY SPRING WHEN OLD AND YOUNG AGE PLANTS START FLOWERING, THE PROTECTIVE MOVEMENTS OF FLOWERS AND LEAVES, STEMS AND INTELLIGENCE SYSTEMS FUNCTIONS, IN COMPARING THE OLD PLANTS, WITH YOUNG PLANTS PERFORMANCES ALL ARE DIFFERENT, DETERIORATE AND CHANGES IN VARIABLE DEGREES AT OLDER PLANTS, MOST FUNCTIONS IN OLDER PLANTS BECOME MUCH SLOW AND WEAKER FUNCTIONS THAN, THE

YOUNGER PLANT, THE YOUNG PLANTS PERFORM EVERY FUNCTIONS ACCURATE, SHARP, AND FAST, IN OLD PLANTS IT LOOKS CLEARLY OLDER PLANTS EVEN MAY FORGET TO DO THEIR JOBS IN MANY FUNCTIONS AS THEY WERE DOING IN YOUNG AGES.

WHEN COMPARING THE SAME SPECIES YOUNG AND OLD PLANTS NEXT TO EACH OTHER IN A FREEZING OR SNOWING EARLY SPRING MORNING, THE GENERAL PROTECTIVE FUNCTIONS OF YOUNG AGE PLANTS ARE FAR BETTER, THE YOUNG PLANTS START CLOSING CUSPIDS WHEN CLIMATE CHANGES TO FREEZING, BUT THE OLDER PLANTS EVEN HAVE NOT STARTED TO DO THE SAME MAY THEIR INTELLIGENCE SYSTEMS ALSO DETERIORATING, FORGETING AND SLOWING DOWN AND NOT FUNCTIONING AT ALL.

EVEN IN SAME AGE PLANTS ALSO ARE A LOT OF DIFFERENT TYPES VARIABLES, THE IMMUNITY SYSTEMS OF SAME SPECIES PLANTS WITH SAME AGES IN ONE PLANT COMPARING TO OTHER, THE FUNCTIONS ALSO ARE VARIED, ONE DO BETTER THAN THE OTHER.

BRAINLESS- PLANTS

SOME PLANTS DO NOT POSSESS INTELLIGENCE SYSTEM CENTER

CONTROL OF STEM CELL'S EXPONENTIAL GROWTH, AND CELL FUNCTIONS, ALL UNDER PARTICLE INTELLIGENCE SYSTEMS:

THERE ARE A LOT OF PLANT SPECIES, WHO DO NOT HAVE INTELLIGENCE SYSTEM CENTERS, AND HAVE NOT DEVELOPED FUNCTIONING PLANT INTELLIGENCE SYSTEMS. THERFORE IN THESE SPECIES, THE DESCRIPTION OF PLANT FUNCTIONS UNDER C. I. S., S.I.S., M.I.S., ARE NOT EXISTED AT ALL. ALSO THERE ARE MANY PLANT SPECIES ONLY HAVE PARTIALLY DEVELOPED INTELLIGENCE SYSTEMS, AND INCOMPLETE INTELLIGENCE SYSTEM CENTERS, THESE SPECIES RESPONSES ARE NOT GOINING TO BE THE SITUATIONS DESCRIBED ABOVE. WHICH WAS ABOUT ADVANCED PLANT SPECIES, WITH FULL DEVELOPED CENTERAL INTELLIGENCE SYSTEM CENTERS.

IN THOSE SPECIES WITH EXISTING C. I. S. IN EARLY SPRING DURING FAST- GROWTH, STEM CELL'S PHASE, WHEN THE NEWLY BORN, EMBRYONIC STEM CELLS START THEIR EXPONENTIAL SPEED MULTIPLICATIONS, AND DUPLICATION PHASE GROWTH, ALL EMBRYONIC PHASES TAKE PLACE UNDER SENSARY REPORTS OF S.I.S. AND ORDERS OF C.I.S., AND THE M.I.S. HAS TO CARRY OUT THE CENTRAL INTELLIGENCE SYSTEMS ORDERS ACCORDINGLY, HOW THE STEM CELLS MUST FUNCTION., AND MUST COORDINATE THEIR FUNCTIONS ACCORDING CELL'S, ATOM'S, INTELLIGENCE SYSTEM'S ORDERS UNDER FUNDAMENTAL PARTICLE OPERATED C. I. S. CENTERS.

AT EARLY STAGE EMBRYONIC PLANT STEM CELL GROWTH PHASES WHEN THE FLOWERS AND BUDS OPENS DURING THE DAY, FOR PROTECTION OF EARLY STAGE PETALS, OVARIES, STAMEN IN COLD CHILLING NIGHTS, AND EARLY MORNING HOURS, UNDER C. I. S. ORDER THE CUSPIDS CLOSE AT NIGHTS, THE EARLY STAGE PETALS, STAMEN AND OVARIES GET PROTECTION FROM CHILLING COLDS DAMAGES OF NIGHT AND EARLY MORNING TIMES.

AND SHORTLY THEREAFTER, WHEN AT THE DAY, THE SUN RISE AND WEATHER GET WARMER, THE S.I.S. SENSE BETTER TEMPERATURE AND AMPLE LIGHT PARTICLES AROUND AND THEIR DIRECTIONS, THE C. I. S. GET THE S.I.S. REPORTS. ISSUE THE ORDERS FOR OPENING CUSPIDS AND BUDS AGAIN, AS THE SUN SHINE WITH AMPLE DIFFERENT LIGHT PARTICLE FROM PLANET SUN FILLS ALL EARTHS AIR, THERE AFTER C. I. S. ORDERS FLOWERS STAY WIDELY OPENS DURING ALL DAY, EVEN THROUGH THOSE C.I.S. ORDERS, STEMS, FLOWERS, LEAVES, TILTS AND ROTATIONS, CHASE THE DIRECTION OF SUNS TRAVEL PATH, EVERY DAY FROM EAST TO WEST, UNDER WELL FUNCTIONING PLANTS CENTRAL INTELLIGENCE SYSTEMS CENTERS ORDER. AND ACCUMULATING THE NEEDED LIGHT PARTICLES FOR CONSTRUCTIONS OF PLANTS ELECTRONS NUCLEONS PARTICLE COMPOUND CONSTRUCTIONS, THERE AFTER LATE AFTERNOON WHEN SUN GOES DOWN TOWARD THE NIGHT AND TEMPERATURE CHANGING TO COLD ONLY IN SOME SPECIES THE FLOWERS AGAIN UNDER CENTRAL PLANT INTELLIGENCE SYSTEM CENTERS START CLOSING CUSPIDS AGAIN, THESE BIOLOGICAL PLANT MOVEMENT CYCLES CONTINUE FOR LIFE WITH NO INTERRUPTION SAME DAY AND NIGHT, THIS IS PHENOMENON OF LIGHT FUNDAMENTAL PARTICLE COLLECTING – STORING CYCLES CONTINUE ONE DAYS AFTER THE OTHER, THE COLLECTED Y. F.P. DIRECTLY INTER INTO SUBSYSTEM-UNITS OF DIFFERENT PLANTS ELECTRONS – NUCLEONS AND PRODUCE ELECTRON NEO-GENESIS, NUCLEON NEO- GENESIS,

ATOM GENESIS OF PLANT CELLS, THESE PHENOMENONS ALREADY EXPLAINED OTHER CHAPTERS AND VOLUMES AS WELL.

THE FLOWERS WHOSE ATOMS CONSTRUCTED WITH HIGH ELECTRIC ENERGY PARTICLES, PROVIDE BETTER IMMUNITY AND DEFENSE SYSTEM TO PLANTS

EFFECTS OF PARTICLE MOLECULAR CONSTRUCTIONS ON PLANT IMMUNITY AND DEFENSE SYSTEMS

==

THE PLANTS CONSTRUCTIONS, THAT HAS BEEN CONSTRUCTED THROUGH USE OF HIGH ENERGY LIGHT PARTICLES COMPOUND CONSTRUCTIONS, AND THOSE WHO HAVE HIGH ELECTRIC ENERGY CONTENT PARTICLE COMPOUNDS, PROVIDE BETTER IMMUNITY AND SURVIVAL TO FLOWERS AND PLANTS.

THE PARTICLES COMPOUND CONSTRUCTED PLANTS, WITH LESS ELECTRIC ENERGIES PARTICLES, PROVIDE LESS IMMUNITY AND DEFENSE CAPABILITIES TO FLOWERS AND PLANTS, THESE PLANT SPECIES CAN NOT DEFEND SELF AGAINST ENVIRONMENTAL DISASTERS, AS THOSE PLANTS, WHO HAVE BEEN CONSTRUCTED WITH HIGH ENERGY PARTICLE COMPOUND CONSTRUCTED PLANTS.

THE IMMUNITY SYSTEMS OF VIOLET COLOR PLANTS ARE BETTER THAN IMMUNITIES SYSTEMS OF WHITE COLOR FLOWERS. THE VIOLET PLANTS POSSESS HIGH ELECTRIC ENERGY VIOLET FUNDAMENTAL PARTICLES.

BETWEEN SENSIBLE LIGHT PARTICLES IN EARTH, THE VIOLET COLOR LIGHT PARTICLES POSSESS HIGH ELECTRIC ENERGY CONTENT THAN, THE OTHER COLOR SENSIBLE PARTICLES. THE FLOWERS WHOSE ELECTRONS AND NUCLEONS CONSTRUCTED WITH VIOLET COLOR LIGHT PARTICLE COMPOUND CONSTRUCTIONS, THOSE FLOWERS POSSESS BETTER AND HIGHER IMMUNITY AND DEFENSIVE SYSTEMS THAN, THE WHITE COLOR FLOWERS, WHOSE INTERNAL ELECTRON NUCLEON PARTICLE COMPOUND CONSTRUCTIONS ARE DONE BY WHITE COLOR PARTICLE COMPOUND CONSTRUCTIONS.

THE FLOWERS THAT HAS BEEN CONSTRUCTED FROM WHITE COLOR LIGHT PARTICLE –COMPOUNDS OR HAVE BEEN CONSTRUCTED FROM YELLOW COLOR LIGHT PARTICLE COMPOUNDS POSSESS LESS ELECTRIC ENERGY CONTENTS, AND THEIR DEFENSIVE SYSTEMS AND IMMUNITY ARE LESS THAN THE FLOWERS WHO HAVE BEEN CONSTRUCTED FROM VIOLET COLOR LIGHT PARTICLE COMPOUND CONSTRUCTIONS, OR VIOLET COLOR FLOWERS.

VIOLET COLOR FLOWER SPECIES IN HARSH DANGEROUS ENVIRONMENTAL CONDITIONS SUCH AS EXTREME COLD OR EXTREME HOT WEATHERS HAVE MUCH MORE CHNACE OF SURVIVAL THAN THE WHITE OR YELLOW COLOR FLOWERS, HAVING MORE ELECTRIC ENERGY CONTENT VIOLET COLOR PARTICLE COMPOUNDS PROVIDE BETTER RESISTENCES AND IMMUNITY AND DEFENSE SYSTEMS AGAINST HOT WEATHER AS WELL AS AGAINST COLD WEATHER HAZARDOUS CONDITIONS, WHICH UNDER THOSE LEVEL HEAT OR FREEZING CONDITIONS MOST WHITE OR YELLOW FLOWERS DIE, BUT THE VIOLET FLOWERS AND PLANTS MAY SURVIVE.

PLACING VIOLET COLOR AND WHITE COLOR PLANTS UNDER COLD ENVIRONMENTS, THE VIOLET COLOR FLOWERS SURVIVE BETTER THAN WHITE COLOR FLOWERS, IT IS THE SAME, PLACING VIOLET AND WHITE COLOR FLOWERS UNDER EXTREME HARSH HOT SUMMER TEMPERATURES WITH DEHYDRATIONS, THE VIOLET COLOR PLANTS POSSESS BETER CHANCES OF SURVIVAL THAN THE WHITE COLOR FLOWERS.

HIERARCHY ORDER SYSTEMS BETWEEN CIS, SIS, MIS IN PLANTS

THE PLANTS SENSARY INTELLIGENCE SYSTEMS CENTERS (S. I. S.) FROM FLOWERS LEAVES STEMS ROOTS SENSE AND REPORT ENVIRONMENTAL CONDITIONS SUCH AS EXISTING LIGHT PARTICLES AT AIR AND ITS DIRECTIONS AS WELL AS THE OUTSIDE TEMPERATURE CONDITIONS EITHER COLD OR WARM TEMPERATURE, EXISTING WATER OR MINERAL CONDITIONS, ETC., AND REPORT ALL OUTSIDE INFORMATIONS THROUGH Y. - F.P. – I.I. – P.cl. TO PLANTS

CENTRAL INTELLIGENCE SYSTEM CENTERS THROUGH PARTICLE CIRCULATION SYSTEMS, UNDER HIERARCHY ORDER SYSTEMS OF REPORTING INFORMATIONS FROM LOWER INTELLIGENCE SYSTEM CENTERS, TO HIGHER INTELLIGENCE SYSTEM CENTERS.

IN CENTRAL INTELLIGENCE SYSTEM CENTERS (C. I. S.) THE INCOMING F.P.- I.I.- P. cl. FROM SENSARY SYSTEMS, INTE INTO PLANTS INTER ELECTRON-NUCLEON SUBSYSTEM – UNITS CHEMICAL LAB.S, COMBINE WITH INTER ELECTRON-NUCLEON PARTICLE COMPOUNDS, AND STORE INFORMATIONS IMAGE AS RECORDS KEEPING SYSTEMS, PRODUCE DIFFERENT PARTICLE –COMPOUNDS CONSTRUCTING OF THE INTERNAL ELECTRON- NUCLEON OF PLANTS C. I. S., THIS IS PHENOMENON OF NEW-PARTICLE-CLOUD COMPOUND CONSTRUCTION PHENOMENONS, AND THAT IS KEEPING AND STORING OUTSIDE INFOMATIONS AND IMAGES RECORDS, AT INSIDE THE PLANTS INTERNAL ELECTRONS AND NUCLEONS IN PARTICLE COMPOUND FORMS.

THE C.I.S. THROUGH A. S. I. F.P. Mol. C.I.C. COMPARE AND ANALIZE INCOMING PARTICLE CLOUDS INFORMATIONS AND IMAGES WITH PRE-EXISTING INTERNAL ELECTRON-NUCLEON PARTICLE COMPOUND INFORMATIONS AND IMAGES, AND THE C.I.S. THEREAFTER MAKE PROPER DECISIONS OVER THESE INTERNAL ELECTRON NUCLEONS INFORMATIONS AND THEIR COMPARISONS WITH EACH OTHER, THROUGH EXPONENTIAL INTERACTION SPEEDS. THEREAFTER THE C.I.S. ISSUE ORDERS TO BE TRANSMITTED TO LOWER HIERARCHY SYSTEMS, BEFORE THE MOTOR INTELLECTUAL SYSTEMS CENTERS, THE ORDERS TO BE EXECUTED ACCORDINGLY AS ISSUED BY C.I.S. EXACTLY. UNDER HIERARCHY SYSTEMS FROM HIGHER CENTERS TO LOWER INTELLIGENCE SYSTEM CENTERS, THE M. I. S. EXECUTE ORDERS AS RECEIVED FROM C. I. S. EXACTLY ACCORDING ORDERS INSTRUCTIONS.

INTRODUCTION TO FUNDAMENTAL PARTICLE'S GENERAL CHEMICATRY

COMBINING CRYSTALLINE INTERNAL ATOM LIGHT PARTICLE COMPOUNDS WITH EXOGENOUS EXTERNAL ATOM FUNDAMENTAL PARTICLES:

CRYSTALLINE ATOMS ARE LIGHT PARTICLE- COMPOUND CONSTRUCTED ATOMS, AND DIFFERENT CRYSTALLINE ELECTRONS –NUCLEONS HAS BEEN CONSTRUCTED FROM DIFFERENT KINDS NUMEROUS LIGHT FUNDAMENTAL PARTICLE COMPOUNDS CONSTRUCTIONS, UNDER REGENERATIVE A. S. I. F.P. Mol. C.I.C. DIFFERENT KINDS EXOGENOUS INCIDENT INCOMING FUNDAMENTAL PARTICLES SUCH AS X- FUNDAMENTAL PARTICLES, ELECTRIC-FUNDAMENTAL PARTICLES, GAMMA-PARTICLES, ETC. ONE BY ONE COMBINED ONE TEST AFTER OTHER, AND EACH GIVEN CHEMICAL COMBINATIONS TEST PRODUCED DIFFERENT KIND PARTICLE BY PRODUCTS (P. B. P.) SHOWED DIFFERENT A. S. I. F.P. Mol. C.I.C. BETWEEN DIFFERENT FUNDAMENTAL PARTICLE MOLECULAR STRUCTURES CAUSE PRODUCTION OF DIFFERENT KIND PARTICLE –BY- PRODUCTS WHICH IN ALL TESTS THE RESULTS ARE A. S. I. F.P. Mol. C.I.C. –SPECIFIC, AND IN DIFFERENT TESTS THE RESULTS ARE DIFFERENT FROM EACH OTHERS.

UNDER FOLLOWING TESTS DIFFERENT INDOGENOUS INTERNAL ELECTRON-NUCLEON LIGHT -PARTICLE COMPOUNDS OF A GIVEN CRYSTALINE COMBINED WITH DIFFERENT EXTERNAL ATOM EXOGENOUS FUNDAMENTAL PARTICLES ONE TEST AFTER OTHER UNDER SEPARATE TESTS AS FOLLOWING:

1- IN FIRST TEST COMBINATION OF GIVEN EXOGENOUS INCIDENT ELECTRIC - PARTICLES WITH GIVEN COMPATIBLE GROUPS OF INTER ELECTRON-NUCLEON LIGHT PARTICLE COMPOUNDS OF GIVEN CRYSTALINE ACHIEVED THROUGH A. S. I. F.P. Mol. C.I.C. UNDER SPECIFIC TEST CONDITIONS AND PRODUCED SPECIFIC GIVEN PARTICLE-BY-PRODUCTS (P.B.P.) SPECIFIC OF MOLECULAR COMBINATIONS.

2 – IN SECOND TEST GIVEN EXOGENOUS INCIDENT X- FUNDAMENTAL PARTICLES COMBINED WITH OTHER GROUPS OF COMPATIBLE GIVEN INTERNAL ELECTRONS –NUCLEONS LIGHT PARTICLE COMPOUNDS OF CRYSTALINE UNDER SPECIFIC TEST CONDITIONS AND PRODUCED SPECIFIC GIVEN PARTICLE –BY – PRODUCTS (P.B.P.) SPECIFIC FOR THIS A. S. I. F.P. Mol. C.I.C.

3 – IN THIRD TEST GIVEN SPECIFIC INCIDENT EXOGENOUS GAMMA FUNDAMENTAL PARTICLES COMBINED WITH DIFFERENT COMPATIBLE INTERNAL ELECTRON –NUCLEON LIGHT – PARTICLE COMPOUNDS OF CRYSTALINE UNDER SPECIFIC TEST CONDITIONS AND PRODUCED TEST SPECIFIC PARTICLE BY PRODUCTS UNDER GIVEN TEST CONDITIONS THROUGH A. S. I. F.P. Mol. C.I.C.

THE ABOVE THREE TESTS, THREE DIFFERENT EXOGENOUS PARTICLES CHEMICALLY COMBINED WITH THREE DIFFERENT KINDS INTERNAL ELECTRON NUCLEONS LIGHT PARTICLE COMPOUNDS, AND PRODUCED THREE DIFFERENT COLOR VISIBLE LIGHT PARTICLES, LIGHT PARTICLES VISIBLY SCAPED AIRBORN.

IT IS OBVIOUS THE CHEMICAL COMBINATION OF GAMMA PARTICLES WITH GIVEN LIGHT –PARTICLE-COMPOUNDS PRODUCE GAMMA-PARTICLE COMPOUND, PLUS A GIVEN SPECIFIC KIND - COLOR LIGHT PARTICLE WHICH THIS RELEASED LIGHT PARTICLE IS SPECIFIC FOR THIS GIVEN TEST AND COMBINING LIGHT PARTILE COMPOUND STRUCTURE, THE RELEASED LIGHT PARTICLE SCAPE FREE AIRBORN AND SENSED WHEN A. S. I. F. P. Mol. C.I.C. ACCOMPLISHED, THE P. B. P. PRODUCED UNDER THIS CHEMICAL COMBINATION IS PRODUCTION OF SPECIFIC COLOR AND KIND LIGHT PARTICLE PLUS GAMMA- PARTICLE COMPOUNDS WHICH ARE USED IN NEW CONSTRUCTIONS OF ELECTRONS- NUCLEONS (PARTICLE –COMPOUND NEO-GENESIS) SPECIFIC OF THIS TEST.

THE A. S. I. F.P. Mol. C.I.C. RESULTS OF ELECTRIC –FUNDAMENTAL PARTICLES COMBINATIONS WITH LIGHT-PARTICLE-COMPOUNDS OF CRYSTALINE PRODUCE PRODUCTION OF ELECTRIC- PARTICLE –COMPOUND PLUS ANOTHER KIND AND COLOR TEST SPECIFIC LIGHT PARTICLE WHICH AGAIN ARE SENSABLE AND SCAPE VISIBLE TO AIR.

THE CHEMICAL COMBINATIONS RESULTS WITH X-PARTICLES PRODUCE DIFFERENT KIND TEST SPECIFIC RESULTS, SOME TESTS MAY RELEASE BRIGHT WHITE LIGHT PARTICLES, THE OTHER RED OR BLUE COLOR LIGHT PARTICLES, ETC.

ABOVE TESTS CAN BE REPEATED WITH REMARKABLY LARGE NUMBERS OF ANOTHER KIND ATOMS OTHER THAT CRYSTALINE OR DIFFERENT KIND OTHER CRYSTALINE ATOMS AND THE ALL TEST RESULTS ALL ARE INTERACTION – SPECIFIC WITH NO CHANGE, EVEN SOME TIMES THE LIGHT PARTICLES WHICH ARE NOT EXIST IN EARTH MAY SCAPE VISIBLE, IN THE SAME TIMES MANY OF THE OTHER LIGHT PARTICLES OR NON-LIGHT PARTICLES OF ANOTHER KINDS OBVIOUSLY ARE NOT DETECTABLE PRESENTLY.

IN THESE TESTS THERE ARE LARGE NUMBERS OF NON- DISCOVERED FUNDAMENTAL PARTICLE ALSO PARTICIPATE AT INTERACTIONS AS WELL AS RELEASED BY THESE TESTS WHICH WE HAVE NO ABILITY TO EVALUATE THOSE AT PRESENT TIME.

ELECTRON-CYCLES

COMBINING AN INTERNAL ELECTRON PARTICLE COMPOUND WITH

INCIDENT EXTERNAL ATOM PARTICLE MOLECULE:

OR PHENOMENON OF TRIGGERING CENTRIFUGAL AND CENTRIPETAL ELECTRON'S TRANSITION CYCLES

SHINE ONE MOLECULAR –UNIT FUNDAMENTAL PARTICLE (PHOTON) INTO A GIVEN ELECTRONS CHEMICAL LAB.S, AND COMBINE EXOGENOUS PHOTON (Y. –F.P.) WITH INTERNAL ELECTRON PARTICLE- COMPOUNDS (F.P. _ COMP.) IN GIVEN ORBIT, COMBINE ONE MOLECULAR UNIT INTER-ELECTRON INDOGENOUS PARTICLE-COMPOUND WITH ONE MOLECULAR-UNIT EXOGENOUS FUNDAMENTAL PARTICLE (Y. –F.P) UNDER A. S. I. F. P. Mol. C.I.C., WILL PRODUCE PHOTON – COMPOUND (Y. F.P. _ COMP.) THIS COMBINATION WILL CAUSE THE WEIGHT OF PARTICLE COMPOUND INCREASE EQUAL TO THE COMBINING PHOTON WEIGHT, EQUATION IS AS FOLLOWING.

Y.- F.P. + F.P. _COMP. ----------→ Y. – F.P. _ COMP.

←-------------

INCREASE IN Y. F.P. _ COMP. WEIGHT, IS EQUAL TO INCREASE IN THE ELECTRON'S WEIGHTS, THESE INCREASES OF ELECTRONS OR PHOTON -COMPOUND WEIGHTS, ARE EQUALS TO THE WEIGHT AMOUNTS OF COMBINING PHOTON THIS A. S. I. F.P. Mol. C.I.C. WILL CAUSE THE ELECTRONS WEIGHTS INCREASE EQUAL TO WEIGHT OF COMBINING PHOTON. INCREASE IN ELECTRON'S WEIGHT IS PROPORTION TO INCREASE IN ELECTRON'S ENERGY AMOUNT, AND ELECTRON'S GRAVITON FORCE INCREASE.

THIS INCREASE IN ELECTRON'S WEIGHT, GRAVITON, ENERGY, WILL CAUSE THE ELECTRON TO TRANSIT FROM LOW ENERGY ELECTRON ORBIT, INTO HIGH ENERGY ELECTRON ORBIT LEVEL, THIS PHENOMENON IS CENTRIFUGAL ELECTRON TRANSIT CYCLE, OR IT IS ELECTRON'S CENTRIFUGAL TRANSITION CYCLE IN DIRECTION OF INDEPENDENT AND ELECTRON TRANSIT TOWARD PERIPHERAL ORBITAL DIRECTIONS.

MOST OF FUNDAMENTAL PARTICLE CHEMICAL COMBINATIONS ARE REVERSIBLE, THAT MEANS ORIGINAL CHEMICAL INTERACTIONS REVERSE ITS ORIGINAL EQUATIONS DIRECTION SPONTANEOUSLY BACKS INTO OPPOSITE DIRECTION OF ORIGINAL REGENERATIVE A.S. I. F.P. Mol. C.I.C.

REGENERATIVE A. S. I. F.P. Mol. C.I.C. CAUSED CONSTRUCTION OF PHOTON PARTICLE COMOUND INCREASE IN ELECTRON WEIGHT, THE DEGENERATIVE A. S. I. F.P. Mol. C.I.C. CAUSE BREAK DOWN OF PHOTON COMPOUND INTO CONSTRUCTING SMALLER MOLECULES, AND RELEASE OF PHOTON BACK AIRBORN, THIS PROCESS CAUSE DECREASE IN WEIGHT OF ELECTRON AND PARTICLE COMOUNDS. IN CONSEQUENCE THE ELECTRONS HAS TO TRANSIT BACK INTO LOW ENERGY LEVEL ORBITS AT REVERSE DIRECTION, THIS PHENOMENON IS CENTRIPETAL ELECTRON- TRANSIT TOWARD DEPENDENCE, OR ITS FINAL TRAVEL INTO COMBINATION WITH NUCLEONS UNDER ATTRACTION GRAVITON FORCES, THIS IS ELECTRON'S CERTRIPETAL TRANSIT CYCLES, THE ELECTRON TRANSIT TOWARD CENTRAL DIRECTIONS.

REGENERATIVE A. S. I. F.P. Mol. C.I.C. PRODUCE GENESIS OF PHOTON _ COMPOUNDS (Y. – F. P. _ COMP.) INSIDE ELECTRONS SUBSYSTEM- UNITS, WHICH CONSTRUCT INTERNAL ELECTRON PHOTON COMPOUND CONSTRUCTIONS OF ELECTRONS, THIS PHENOMENON IS NEO- GENESIS OF FUNDAMENTAL PARTICLE COMPOUNDS, CAUSING NEO-GENESIS OF ELECTRONS.

PRODUCTION OF AIRBORN LIGHT PARTICLE CLOUDS, OR PARTICLE CLOUD GENESIS

VISIBLE PARTICLE CLOUDS GENESIS

COMBINING AIRBORN EXOGENOUS LIGHT PARTICLES, WITH INDIGENOUS LIGHT PARTICLE COMPOUNDS

H – O – H ATOMS ARE CONSTRUCTED FROM LIGHT FUNDAMENTAL PARTICLE - COMPOUND CONSTRUCTIONS, THE H – O – H OR WATER ATOMS INTERNAL ELECTRON, NUCLEON PARTICLE COMPOUNDS HAVE BEEN CONSTRUCTED FROM LARGE QUANTITIES OF RED, YELLOW, GREEN, BLUE, VIOLET, LIGHT PARTICLE COMPOUNDS CONSTRUCTIONS.

THE INTERNAL ELECTRON, NUCLEON INDIGENOUS DIFFERENT LIGHT PARTICLE COMPOUNDS CONSTRUCTIONS OF WATER ATOMS, UNDER THE A. S. I. F.P. Mol. C.I.C. COMBINE WITH EXTERNAL ELECTRON- NUCLEON EXOGENOUS PARTICLES, SUCH AS WHITE LIGHT PARTICLES, AFTER COMBINATION THE RELEASED, FREE LIGHT PARTICLES IN FORMS OF AIRBORN PARTICLE CLOUDS SCAPE INTO AIR, AS RED, YELLOW, GREEN, BLUE, VIOLET DIFFERENT COLOR LIGHT PARTICLES CLOUDS AGAIN.

PARTICLE CLOUD GENESIS

IN WARM TEMPERATURE SUMMER CONDITIONS, SHINE WHITE COLOR LIGHT FUNDAMENTAL PARTICLES AIRBORN INTO ELECTRONS, NUCLEONS OF VAPORIZED H – O – H AIRBORN ATOMS AT AIR. AND COMBINE WHITE INCOMING EXOGENOUS LIGHT PARTICLES FROM AIR WITH INTERNAL ELECTRON NUCLEON DIFFERENT LIGHT PARTICLE COMPOUNDS OF DIFFERENT COLORS, SUCH AS: RED LIGHT PARTICLE COMPOUNDS, YELLOW AND GREEN LIGHT PARTICLE COMPOUNDS, OR COMBINE WITH INTERNAL ELECTRON NUCEON BLUE AND VIOLET DIFFERENT LIGHT PARTICLE COMPOUNDS, UNDER REGENERATIVE A. S. I. F.P. Mol. C.I.C.

THESE MULTI- A.S.I. P.F. Mol. C.I.C. BETWEEN EXOGENOUS WHITE LIGHT PARTICLE COMBINATION WITH INDIGENOUS DIFFERENT PARTICLE COMPOUNDS COMBINATIONS, WILL PRODUCE DIFFERENT COLOR LIGHT PARTICLES, WHICH THESE RELEASED LIGHT PARTICLES SCAPE INTERNAL ELECTRON, NUCLEON REGIONS BACK TO AIR, THE PRODUCED PARTICLE CLOUD AT AIR ARE VISIBLE AS: RED LIGHT PARTICLE CLOUDS. YELLOW LIGHT

PARTICLE CLOUD, GREEN PARTICLE CLOUDS, BLUE AND VIOLET LIGHT PARTICLE CLOUDS. AND DECIPATE AIRBORN TO SPACE SHORTLY AFTER PRODUCTION AND DISAPPEAR.

FROM OTHER A. S. I. F.P. Mol. C.I.C. OCCURRENCES IN COMBINATIONS OF AIRBORN BIOFRIENDLY WHITE COMPOSITE LIGHT PARTICLES, WHICH THESE EXOGENOUS WHITE LIGHT PARTICLE HAVE BEEN COMPOSED FROM VARIOUS DIFFERENT CLASSES OF DIFFERENT COLOR LIGHT FUNDAMENTAL PARTICLES, AND DURING THE A. S. I. F.P. Mol. C.I.C. SOME OF THESE WHITE FUNDAMENTAL PARTICLES COMPOSITE COMPONENTS, DIRECTLY INTER INTO DIFFERENT H – O – H INTERNAL ELECTRON NUCLEON SUBSYSTEM UNITS CHEMICAL LAB.S AND COMBINE WITH INTERNAL ELECTRON NUCLEON PARTICLE COMPOUNDS OF H – O- H ATOMS, AND THE OTHER REMAINING LIGHT FINADAMENTAL PARTICLES HAS TO SCAPE INTERNAL ATOM SPACE, TO THE OUT SIDE INTO AIR, AND THESE FREE MULTI- CLASS PARTICLES ARE VISIBLE AS VISIBLE LIGHT PARTICLE-CLOUDS, STANDING WITH EACH OTHER UNDER SPECIFIC PARTICLES LAWS AND RULES. WHICH WE CAN SEE THEM AS VISIBLE PARTICLE CLOUDS IN AIR.THIS IS PHENOMENON OF LIGHT PARTICLE CLOUD – GENESIS, UNDER A. S. I. F.P. Mol. C.I.C.

ALL A. S. I. F.P. Mol. C.I.C. ARE FAST WITH EXPONENTIAL SPEEDS RATE OF CHEMICAL INTERACTIONS, ABOVE AUTONOMOUS SEQUENTIAL CHEMICAL INTERACTIONS LASO ARE WITH NO EXCEPTIONS, THE ABOVE A.S.I. F.P. Mol. C.I.C. WILL IMMEDIATELY CEASE, IF WE STOP H-O-H VEPORIZATIONS, ALSO IT WILL STOP, IF WE CEASE THE SHINING OF EXOGENOUS LIGHT PARTICLES, BOTH WILL STOP THE PRODUCTIONS OF PARTICLE CLOUDS SIMILAR TO "ON – OFF" ELECTRIC KEY INSTANTLY. THIS IS SIMPLE TEST TO DEMONSTRATE EXISTING PARTICLE CLOUD VISIBLE TO

NAKED EYES. THIS PHENOMENON IS GENESIS OF LIGHT FUNDAMENTAL PARTICLE CLOUDS, THROUGH A.S.I. F.P. Mol. C.I.C.

PRODUCTION OF, VISIBLE LIGHT FUNDAMENTAL PARTICLE –CLOUDS, IN NATURE, HIGH ON SKY

THIS PHENOMENON IS A NATURAL OCCURRING A. S. I. F.P. Mol. C.I.C. TAKING PLACE HIGH IN SKY, BETWEEN COMBINATION OF AIRBORN, EXOGENOUS, SUN ORIGIN, INCOMING, WHITE COLOR LIGHT FUNDAMENTAL PARTICLE WHICH COMBINING WITH H – O – H INTERNAL ELECTRON NUCLEON, MULTIPLE DIFFERENT LIGHT PARTICLE COMPOUNDS, WHICH THESE A. S. I. F.P. Mol. C.I.C. TAKING PLACE HIGH IN CLOUDS OF THE SKY, (CLOUDS ARE H. – O – H. ATOMS), THESE A. S. I. F.P. Mol. C.I.C PRODUCE MULTIPLE DIFFERENT COLOR LIGHT PARTICLE CLOUDS OF DIFFERENT COLORS SUCH AS: RED, YELLOW, GREEN, BLUE AND VIOLET COLORS LIGHT FUNDAMENTAL PARTICLE CLOUDS, HIGH IN THE SKY. WHICH THESE MULTI COLOR LIGHT PARTICLE CLOUDS ARE VISIBLE FROM MILES AWAY FROM EARTH, AND THESE AIRBORN LIGHT PARTICLE CLOUDS ARE CALLED RAIN BOW BY PUBLIC. RAIN BOW PRODUCTION IS EXACTLY THE SAME A. S. I. F.P. Mol. C.I.C. AS EXPLAINED IN PREVIOUS PAGES AND ABOVE.

PHENOMENON OF PARTICLE – CLOUD GENESIS

GENESIS OF AIRBORN PARTICLE CLOUDS

COMBINING EXOGENOUS LIGHT PARTICLES WITH CRYSTALINE INTERNAL ELECTRONS-NUCLEONS INDIGENOUS LIGHT-PARTICLE COMPOUNDS:

CRYSTALINE ELECTRONS NUCLEONS ARE LIGHT-PARTICLE CONSTRUCTED STRUCTURES AT MOST, THE WATER (H – O – H ATOMS), AND CRYSTALINE ATOMS, BOTH ARE LIGHT FUNDAMENTAL PARTICLE CONSTRUCTED ATOMS, AND THEIR INTERNAL ELECTRON-NUCLEON CONSTRUCTIONS HAVE BEEN CONTRUCTED FROM DIFFERENT KIND OF LIGHT FUNDAMENTAL PARTICLE COMPOUNDS, THE DIFFERENT NON VISIBLE LIGHT PARTICLE CLASSES, AND

DIFFERENT VISIBLE LIGHT FUNDAMENTAL PARTICLE COMPOUNDS SUCH AS RED LIGHT PARTICLE COMPOUNDS, GREEN LIGHT PARTICLE COMPOUNDS, BLUE OR VIOLET COLOR LIGHT PARTICLE COMPOUNDS OR YELLOW COLOR LIGHT PARTICLE COMPOUNDS CONSTRUCT DIFFERENT ELECTRON-NUCLEOM PARTICLE COMPOUND CONSTRUCTIONS OF CRYSTALINE ATOMS,

IN A TEST, SHINE AIRBORN EXOGENOUS WHITE LIGHT FUNDAMENTAL PARTICLES INTO CRYSTALINE INTERNAL ELECTRON NUCLEON SUBSYSTEM CHEMICAL LAB.S, AND COMBINE EXOGENOUS WHITE LIGHT PARTICLES WITH INDIGENOUS INTERNAL ELECTRON NUCLEON DIFFERENT COLOR LIGHT PARTICLE COMPOUNDS, THROUGH A. S. I. F.P. Mol. C.I.C., THE CHEMICAL INTERACTIONS BETWEEN EXOGENOUS INCOMING WHITE LIGHT FUNDAMENTAL PARTICLES, AND INDIGENOUS CRYSTALINE INTER ELECTRON-NUCLEON DIFFERENT COLOR INTER ELECTRON – NUCLEON LIGHT PARTICLE COMPOUNDS WILL PRODUCE DIFFERENT COLOR RELEASED FREE LIGHT FUNDAMENTAL PARTICLES, AS WELL AS WHITE- LIGHT –PARTICLES –COMPOUNDS, AS PARTICLE BY PRODUCTS.

THROUGH THESE CHEMICAL INTERACTIONS, THE INTERACTION PRODUCED AND RELEASED DIFFERENT COLOR FREE LIGHT FUNDAMENTAL PARTICLE CLOUDS, OF DIFFERENT COLORS SUCH AS RED, YELLOW, GREEN, BLUE, VIOLET, ETC. COLOR LIGHT FUNDAMENTAL PARTICLE CLOUDS, WILL BE VISIBLE AT THE EXIT SIDE OF CRYSTALINE, AS MULTI COLOR DIFFERENT KINDS LIGHT PARTICLE CLOUDS, WHICH EACH COLOR IS REPRESENTATIVE OF ONE CLASS LIGHT FUNDAMENTAL PARTICLE CLOUD, ARE DETECTABLE IN PARTICLE CLOUD THROUGH NAKED EYES, SHORTLY AFTER PRODUCTION WILL DECIPATE INTO AIR, AS FREE PARTICLES DISAPPEAR IN SPACE.

THESE PRODUCED FREE AIRBORN DIFFERENT COLOR LIGHT FUNDAMENTAL PARTICLE CLOUDS, AND EACH COLOR PARTICLE CLOUD IS REPRESENTAVIES OF THEIR COLOR SPECIFIC CLASS LIGHT PARTICLE CLASSES. THESE PARTICLE CLASS PARTICLE CLOUDS WILL STAND ORDERLY UNDER PARTICLE ORDERS WITH EACH OTHER, SUCH AS: RED, YELLOW, GREEN, BLUE, VIOLET COLOR LIGHT PARTICLE CLOUDS IN SPECTRAL LINE, AND AFTER TERMINATIONS AND CEASE OF A. S. I. F.P. Mol. C.I.C., THE INDIVIDUALLY PRODUCED LIGHT PARTICLE-CLOUDS OF DIFFERENT COLORS, WILL DECIPATE IN STANTLY AND TRAVEL IN SPACE TOWARD IT'S DESTINATIONS. FOLLOWING ARE SOME DEMONSTRATIONS FROM INTERACTIONS:

Y. w. – F.P. + Y. r. – F.P. _ COMP. ----------→ Y. w. – F.P. _ COMP. + Y. r. –F.P.

←------------

Y. w. – F.P. + Y. y. –F.P. _ COMP. ----------------→ Y. w. – F.P. _ COMP. + Y. y. – F.P.

←----------------

Y. w. – F.P. + Y. g. –F.P. _ COMP. -----------------------→ Y. w. F.P. _ COMP. + Y. g. F.P.

←---------------------------

Y. w. – F.P. + Y. b. – F.P. _ COMP. -------------------→ Y. w. – F.P. _ COMP. + Y. b. –F.P.

←-------------------

Y. w. – F.P. + Y. v. – F.P. _COMP. ----------------------→ Y. w. – F.P. _ COMP. + Y. v. –F.P.

←-------------------

THE INTELLIGENT ELECTRON GENESIS,

AND

INTELLIGENT NUCLEON GENESIS

GENESIS OF INTELLIGENT ELECTRONS, NUCLEONS, PRODUCED, THROUGH USE OF S. Y. – F.P. – I.I. – P. cl. TO CONSTRUCT CNS, ELECTRONS – NUCLEONS, PARTICLE COMPOUND CONSTRUCTIONS, THROUGH THE BUILDING ELECTRONS NULEONS PARTICLE COMPOUND CONSTRUCTIONS BY USE OF SONIC –LIGHT PARTICLE CLOUDS AND CONSTRUCTIONS OF INTELLIGENT S. Y. –F.P. – I.I.- P.cl. _COMPOUNDS, INSIDE CNS ELECTRONS AND NUCLEONS, WHICH IT IS ORIGIN OF THOUGHT CURRENT SYSTEMSS.

UNDER REGENERATIVE A. S. I. F.P. Mol. C.I.C. COMBINE EXOGENOUS S. Y.- F.P. – I.I. – P. cl. WITH CNS INTER ELECTRON- NUCLEON PARTICLEE- COMPOUNDS AND PRODUCE INTERNAL ELECTRON-NUCLEON F.P. – I.I. – P.cl. _ COMPOUNDS.

ENVIRONMENTAL EXISTING THINGS PRODUCED S. Y. - F.P. – I.I. – P. cl. AIRBORN TRANSIT INTO RECIPIENTS CNS INTER CNS ELECTRONS –NUCLEONS, COMBINE WITH INTER ELECTRON-NUCLEON PARTICLE COMPOUNDS INSIDE SUBSYSTEM –UNITS CHEMICAL LAB.S, UNDER A. S. I. F.P. Mol. C.I.C., AND PRODUCE S. Y. - F.P. – I.I. – P.cl. __COMPOUNDS, THESE VIRTUAL PARTICLE CLOUD COMPOUNDS CONSTRUCT CNS ELECTRONS – NUCLEONS CONSTRUCTIONS,

THIS IS STORAGE AND ACCUMULATIONS OF ENVIRONMENT SUBJECTS VIRTUAL PARTICLE CLOUDS, ABOUT ALL KIN SOCIAL POLITICAL FINANCIAL EDUCATIONAL OR SCIENCE MATTERS WHICH STORES INSIDE BRAINS ELECTRONS AND NUCLEONS IN PARTICLE –COMPOUN FORMS, LIVING THINGS PILE UP OF DIFFERENT KINDS VIRTUAL PARTICLE CLOUDS COMPOUNDS, INSIDE THEIR BRAINS, AND CONSTRUCT THEIR CNS ELECTRONS –NUCLEONS WITH S.Y. – F.P –I.I. – P.cl. THROUGH BUILDING ELECTRONS NUCLEON CONSTRUCTIONS WITH USE OF VIRTUAL PARTICLE CLOUDS COMPOUNDS, AND THIS IS PILING UP OF DIFFERENT KINDS KNOWLEDGE- INTELLIGENCE –SCIENCE ETC.IN BRAIN LIFE LONG, IT IS PHENOMENON OF LEARNING GAINING KNOWLEDGE ABOUT DIFFERENT MATTERS, ABOVE PHENOMENON IS STORAGE OF S.Y. – F.P. – I.I.- P.cl. INSIDE THE CNS ELECTRONS NUCLEONS IN PARTICLE CLOUD COMPOUND FORMS.

COMBINING EXOGENOUS S. Y. - F.P. – I.I. – P. cl. WITH INTER-ELECTRON-NUCLEON F. P. _ COMPOUNDS

OR S.Y. – F.P. – I.I.- P. cl. STORAGE INSIDE ELECTRONS NUCLEONS

AND S. Y. –F.P. – I.I. – P. cl. RETRIEVAL FROM INTER – ATOM PARTICLE COMPOUND TO OUSIDE

PARTICLE CLOUD STORAGE, AND PARTICLE CLOUD RETRIEVAL, FROM ELECTRONS NUCLEONS

THIS CHEMICAL COMBINATION TAKE PLACE BETWEEN GIVEN EXTERNAL ATOM EXOGENOUS SONIC-LIGHT FUNDAMENTAL PARTICLE INFORMATION- IMAGE PARTICLE CLOUDS (S. Y. – F.P. – I.I. – P. cl.) COMBINING WITH CNS INTERNAL ELECTRON –NUCLEON PARTICLE COMPOUNDS, AND COMBINATION PRODUCE S. Y. – F.P. – I.I.- P. cl. _ COMPOUND (OR VIRTUAL PARTICLE CLOUD COMPOUND) INSIDE CNS RECIPIENT ELECTRONS- NUCLEONS, OR INSIDE NON-LIVE ATOMS SUCH AS SILICON ELECTRONS-NUCLEONS WHEN COMBINING WITH NON-LIVE ELECTRONS-NUCLEONS.

THESE CHEMICAL COMBINATIONS PRODUCE INTELLECTUAL VIRTUAL PARTICLE CLOUD COMPOUNDS CONSTRUCTIONS, AND CONSTRUCT RECIPIENT'S CNS ELECTRONS –NUCLEONS PARTICLE COMPOUND CONSTRUCTIONS, THROUGH COMBINATIONS OF VIRTUAL – PARTICLE CLOUDS (INTELLECTUAL S. Y. – F.P. – I.I. – P. cl. MOLECULAR STRUCTURES) WITH CNS INTERNAL ATOM PARTICLE COMPOUNDS,

THESE S. Y. – F.P. –I.I. –P. cl. CONTAIN AUDIO-VISUAL INFORMATIONS IMAGES KNOWLEDGE ABOUT OUTSIDE WORLDS ENVIRONMENTAL SUBJECTS, THESE INFORMATION IMAGE PARTICLE –CLOUDS ARE PARTICLE –CLOUD COPIES OF ORIGINAL SUBJECTS, THE VIRTUAL CLOUDS ARE EXACT COPIES OF ORIGINAL SUBJECTS PARTICLE COPIES, AND MADE IN PARTICLE – CLOUD –FORMS.

WHEN AIRBORN VIRTUAL PARTICLE CLOUDS INTER INTO BRAINS AND CONSTRUCT S. Y. –F.P. – I.I. – P. cl. _ COMPOUND INSIDE RECIPIENT CNS ELECTRON-NUCLEON, THIS PHENOMENON IS EQUAL TO STORING ORIGINAL SUBGECTS VIRTUAL PARTICLE COPIES, INSIDE CNS ELECTRONS NUCLEONS AND ATOMS, WHEN THESE VIRTUAL PARTICLE CLOUDS MOVES IT FEEL EXACTLY THE ORIGINAL SUBJECTS AND ORIGINAL MATTERS MOVED INSIDE THEIR CNS ELECTRONS NUCLEONS, THIS IS THE BIOLOGICAL FEELINGS THAT PRODUCE ANIMALS THOUGHT CURRENTS AND PSYCHE.

THE VIRTUAL –PARTICLE CLOUDS ACTIONS ALL ARE BIOLOGICALLY SENSED AS IDENTICAL TO ORIGINAL SUBJECTS ACTS AND FUNCTIONS, THIS IS PHENOMENON OF PSYCHE –GENESIS AND THOUGHT CURRENT –GENESIS PHENOMENON, WHICH VIRTUAL PARTICLE CLOUD IS IDENTICAAL TO THE REAL ORIGINAL SUBJECTS AND MATTERS, BIOLOGICALLY LIVING THINGS SENSE ALL OF THESE PHENOMENONS INSIDE THEIR BRAINS AND KALL IT THE SOUL.

THEREFORE S.Y. –F.P. – I.I. – P. cl. COMPOUNDS (VIRTUAL –CLOUDS COMPOUNDS) STORE DIFFERENT ENVIRONMENTAL AUDIO-VISUAL INFORMATIONS IMAGES, KNOWLEDGES ABOUT DIFFERENT OUT SIDE EXISTING THINGS AND MATTERS INSIDE CNS ELECTRONS – NUCLEONS.

THE CONSTRUCTIONS OF CNS ELECTRONS- NUCLEONS WITH S. Y. – F.P. – I.I. – P. cl. COMPOUNDS IN REALITY ARE SIMILAR TO A LIBERARY OR ENCYCLOPEDIA EVERY THING IS STORED THERE AT INSIDE YOU CAN FIND RELATIVELY.

THESE S. Y.- F.P. –I.I. – P.cl. ALSO CAN BE RELEASED FREE AND RETRIEVED TO OUT OF S. Y. – F.P. I.I. – P. cl. _ COMPOUNDS, THROUGH DEGENERATIVE A. S. I. F.P. Mol. C.I.C., WHICH THIS IS RETRIEVAL PHENOMENON OF PARTICLE CLOUDS TO OUSIDE WORLD.

S. Y. – F.P. – I.I. – P. cl. STORAGE PHENOMENON INSIDE ELECTRONS-NUCLEONS

STORAGE OF PARTICLE CLOUDS INSIDE ELECTRONS -NUCLEONS,

BRIEFLY EXOGENOUS EXTERNAL S. Y. – F.P. – I.I. – P. cl. (VIRTUAL CLOUDS) THROUGH SENSARY PARTICLE TRANSMISSION ROUTES INTER INTO DIFFERENT CNS SENSARY CENTERS ELECTRONS NUCLEONS, UNDER A.S. I. F.P. Mol. C.I.C. COMBINE WITH INTERNAL ELECTRON- NUCLEON PARTICLE COMPOUNDS, (UNDER REGENERATIVE AUTONOMOUS SEQUENTIAL IN PARTICLE – CLOUD CHEMICAL INTERACTION CYCLES), AND COMBINATION PRODUC S. Y. - F.P. – I.I. – P.cl. _ COMPOUND (VIRTUAL- CLOUD COMPOUND), AND STORE ENVIRONMENTAL EVENTS INFORMATIONS IMAGES IN S. Y. - F.P. – I.I. – P.cl. – COMPOUNDS FORMS INSIDE ELECTRONS- NUCLEONS OF RECIPIENT ATOMS, SUCH AS NON- LIVE SILICON ATOMS IN NON-LIVE WORLD OF ATOMS, OR IT STORE INTELLIGENT MOLECULAR S. Y. – F.P. – I.I. – P. cl. – COMPOUNDS (VIRTUAL CLOUD COMPOUNDS) THE INFORMATION AND ENVIRONMENTAL KNOWLEDGES INIDE CNS ELECTRONS NUCLEONS LIVING THINGS SPECIES.

THIS PHENOMENON IS RECORDING OR STORING OR PROGRAMMING OF OUTSIDE ENVIRONMENTAL INCIDENTS INFORMATIONS- IMAGES INSIDE ELECTRONS- NUCLEONS, OR SIMPLY RECORDING KNOWLEDGE INSIDE ELECTRONS-NUCLEONS LIVING THINGS SUCH AS ANIMALS OR INSIDE NON LIVE ATOMS SUCH AS SEMI-CONDUCTORS ATOMS.

ABOVE IS LEARNING PHENOMENON AND KNOWLEDGE – ACCUMULATION PROCESS, AND ARE USED IN HOME BY PARENTS, OR AT SCHOOLS TO TEACH HARD CORE SCIENCE OR ANY THING ELSE IN CLASS ROOMS FROM KINDER GARDEN UP TO THERE AFTER POSTGRADUATE TEACHING SYSTEMS OF DIFFERENT UNIVERSITIES, OR AT DIFFERENT TRAINING FILEDS AND EDUCATIONAL PROGRAMS STARTS FROM KINDER GARDENS AND CONTINUE FOR ALL LIFE, THESE USED IN SOCIAL POLITICAL RELIGIOUS OR ANY OTHER FILED ETC. FOR LEARNING AND ACCUMULATIONS OF ALL DIFFERENT LEGAL OR ILLEGAL TRAININGS.

THESE STORED KNOWLEDGES ALL HAVE ONE HALF LIFE TIME MOLECULAR CONSTRUCTION STABILITIES, AFTER EACH ONE HALF LIFE TIME THEIR QUANTITIES DECLINE APPROXIMATELY TO ONE HALF OF ACCUMULATE QUANTITY, THESE STORAGE S.Y. –F.P.- I.I. – P. cl. EVEN AFTER DEMISE OF LIVING THINGS STAY INSIDE ELECTRONS –NUCLEONS ACCORDING THEIR HALF LIFE SCHEDULES AND GRADUALLY DISINTEGRATE UNDER REVERSE A. S. I. F.P. Mol. C.I.C.

RELEASE AND RETRIEVAL OF PARTICLE – CLOUDS TO OUT OF PARTICLE -COMPOUNDS

PHENOMENON OF RETRIEVAL OF S. Y. – F.P. I.I. – P. cl. TO OUTSIDE CNS ELECTRONS-NUCLEONS

THESE STORED S. Y. - F.P. – I.I. – P.cl. – KNOWLEDGES IN PARTICLE –CLOUD MOLEULAR FORMS AT LATER TIMES CAN BE RELEASED TO OUT OF ELECTRONS- NUCLEONS UNDER DEGENERATIVE A. S. I. F.P. Mol. C.I.C. AS FREE RELEASED PARTICLE –CLOUDS AND THEIR INFORMATIONS- IMAGES OR THEIR MOLECULAR KNOWLEDGES CAN BE DEMONSTRATED AND VISUALIZED ON T.V. SCREENS AGAINS IN AUDIO-VISUAL FORMS.

INTRODUCTION TO NANO – UNIT CHEMISTRY

NANO-UNIT CHEMISTRY AND CHEMICAL COMBINATIONS BETWEEN NANO-UNITS

ELECTRON-GENESIS, PROTON-GENESIS, NEUTRON-GENESIS, AND ATOM – GENESIS

THE CHEMICAL COMBINATIONS BETWEEN THE ELECTRONS AND NUCLEONS

THE NANO-UNITS, CHARACTORS OF THE NANO-UNITS.

DURING MOLECULAR EVOLUTION A. S. I. F.P. Mol. C.I.C. BETWEEN DIFFERENT FUNDAMENTAL PARTICLES AND PARTICLE COMPOUNDS CONSTRUCTED PRIMARY NANO-UNITS ORGANS AND SYSTEMS SUCH AS SUBSYSTEM – UNITS CHEMICAL LAB.S, PARTICLE TRANSPORTATIONS SYSTEMS, SUBSYSTEMS – UNITS INTELLIGENCE SYSTEM CENTERS, ETC., ALSO CONTINUATION OF DIFFERENT OTHER CHEMICAL COMBINATIONS BETWEEN PRODUCED NANO-UNITS PRODUCED MORE DIFFERENT QUANTUM SIZE ANOTHER CLASS STRUCTURES UNDER AUTONOMOUS SEQUENTIAL INTER NANO-UNIT CHEMICAL INTERACTION CYCLES, GRADUALLY DIFFERENT NANO-UNITS SUCH AS ELECTRONS, POSITRONS, QUARKS OR SYSTEMS AND SUBSYSTEM STRUCTURES, PROTONS, NEUTRONS, ATOMS, BARYONS, LEPTONS, MESONS, ETC. ONE AFTER OTHER PRODUCED, WHICH ALL ARE DIFFERENT SIZES DIFFERENT CLASSES NANO- UNITS THEIR FUNDAMENTAL TOTAL POPULATIONS FROM ONE NANO-UNIT TO OTHER VARIES REMARKABLY.

TRANSFORMATION OF NANO-UNITS AND MUTATIONS OF NANO-UNITS FROM ONE TO OTHER THOSE ARE ALSO ANOTHER SUBJECTS WHICH OCCUR FREQUENTLY, HERE MOSTLY DIFFERENT AUTONOMOUS SEQUENTIAL INTER NANO-UNIT CHEMICAL INTERACTIONS CYCLES AND CHAINS (A. S. I. N.-U. C.I.C.) BETWEEN DIFFERENT NANO-UNITS BRIEFLY EXPLAINED AS FOLLOWING.

REGARDING PHYSIOLOGICAL FUNCTIONS IN NANO-UNITS AND FUNDAMENTAL PARTICLES, THE INTER-PARTICLE BEHAVIORS SUCH AS HOSTILITIES OF ONE PARTICLE TO OTHER, OR COOPERATION BETWEEN PARTICLES, THAT ONE PARTICLE DAMAGE ACTIVITIES OF OTHER, AND MAKE OTHER PARTICLE MALFUNCTION, OR PARTICLES HELP EACH OTHER AND COOPERATE, ETC. NOW PARTICLES ARE USED IN CYBER-ATTACKS OF ONE PARTICLE AGAINST OTHERS, ONE PARTICLE DISABLE THE OTHER, THESE ARE NEW WEAPONS.

PROTON-NEO-GENESIS, PROTON-GENESIS NANO-UNIT CHEMISTRY

THREE EXAMPLES FROM THE INTER- NANO-UNIT INTERACTIONS AS FOLLOWING PRODUCE PROTONS:

1 - UNDER DEGENERATIVE AUTONOMOUS SEQUENTIAL INTER NANO-UNITS CHEMICAL INTERACTION CYCLES AND CHAINS (A. S. I. N. – U. C.I.C.): FROM CHEMICAL CONSTRUCTION OF ONE NEUTRON, REMOVE ONE ELECTRON AND ONE ELECTRON ANTI-NEUTRINO, WILL PRODUCE ONE PROTON. THIS IS PROTON-GENESIS UNDER A. S. I. N.-U. C.I.C.:

NEUTRON - ELECTRON - Electron Antineutrino ---------→ PROTON

←------------

2 – UNDER REGENERATIVE A. S. I. N. – U. C.I.C.: COMBINE ONE POSITRON AND ONE ELECTRON NEUTRNO, WITH ONE NEUTRON, WILL PRODUCE ONE PROTON. THIS IS PROCESS OF PROTON –GENESIS, UNDER A. S. I. N. – U. C.I.C.:

NEUTRON + POSITRON + Electron Neutrino ------------→ PROTON

←---------------

3 - UNDER DEGENERATIVE AND REGENERATIVE AUTONOMOUS SEQUENTIAL INTER NANO-UNIT CHEMICAL INTERACTION CYCLES: FROM ONE NEUTRON CONSTRUCTION, REMOVE ONE ELECTRON, ALSO ADD ONE ELECTRON-NEUTRINO, WILL PRODUCE ONE PROTON.

NEUTRON - ELECTRON + Electron Neutrino ----------→ PROTON

←-------- ---

THE DIFFERENT NEUTRONS, PROTONS, ELECTRONS, POSITRONS, NEUTRINO, ANTINEUTRINO INTERACTIONS WITH EACH OTHER THROUGH THE SEQUENTIAL AUTONOMOUS INTER NANO-UNIT CHEMICAL INTERACTIONS UNDER FUNDAMENTAL-PARTICLE EVOLUTION ORDERS AND PATHS CAUSE PRODUCTIONS OF ANOTHER NANO-UNITS, DIFFERENT CHEMICAL INTERACTIONS BETWEEN DIFFERENT NANO-UNITS CREATE NEW PROTONS, NEUTRONS, ATOMS AND ELECTRONS, ETC., THESE ARE PHENOMENONS OF NANO-UNIT NEO- GENESIS.

NEUTRON GENESIS
NEUTRON NEO – GENESIS
(NANO-UNITS CHEMISTRY)

THE FOLLOWING THREE NEUTRON –GENESIS INTERACTIONS, UNDER A. S. I. N.- U. C.I.C. PRODUCE NEUTRON:

1 - UNDER DEGENERATIVE AUTONOMOUS SEQUENTIAL INTER NANO –UNIT CHEMICAL INTERACTION CYCLES AND CHAINS: REMOVE ONE POSITRON, AND ONE NEUTRINO, FROM ONE PROTON CONSTRUCTION, WILL PRODUCE ONE NEUTRON. THIS IS NEUTRON GENESIS PROCESS.

PROTON - POSITRON - Electron Neutrino ----------→ NEUTRON

←------------

2 - THROUGH REGENERATIVE INTER-NANO-UNIT AUTONOMOUS SEQUENTIAL CHEMICAL INTERACTIONS: COMBINE ONE PROTON WITH ONE ELECTRON, AT SAME TIME REMOVE ONE ELECTRON ANTI-NEUTRINO, WILL PRODUCE ONE NEUTRON. UNDER DEGENERATIVE INTER NANO-UNIT CHEMICAL INTERACTIONS. THIS IS NEUTRON GENESIS:

PROTON + ELECTRON - Electron Antineutrino ------------→ NEUTRON

←--------------

3 - UNDER REGENERATIVE AUTONOMOUS SEQUENTIAL INTER-NANO-UNIT CHEMICAL INTERACTIONS:

COMBINE ONE PROTON WITH ONE ELECTRON, AND REMOVE ONE ELECTRON NEUTRINO, WILL PRODUCE NEUTRON. THROUGH DEGENERATIVE A. S. I. N.- U. C.I.C. AND THIS IS PHENOMENON OF NEUTRON GENESIS.

PROTON + ELECTRON - Electron-Neutrino ---------------→ NEUTRON

←-------------------

NANO-UNIT CHEMISTRY

ELECTRON-GENESIS, POSITRON-GENESIS

==

FOLLOWING ARE COUPLE EXAMPLES FROM PROCESS OF ELECTRON GENESIS, POSITRON GENESIS,

UNDER A. S. I. N. – U. –C.I.C.:

1 - ELECTRON- GENESIS:

UNDER DEGENERATIVE – REGENERATIVE A. S. I. N. –U. C.I.C. FROM ONE NEUTRON CONSTRUCTION REMOVE ONE PROTON, AND ADD ONE ANTI-NEUTRINO, WILL PRODUCE ONE ELECTRON.THIS IS PHENOMENON OF ELECTRON-GENESIS.

NEUTRON - PROTON + Electron Anti-Neutrino -----------------→ ELECTRON

←---------------

2 – ELECTRON – GENESIS:

UNDER NUMEROUS DEGENERATIVE AND REGENERATIVE AUTONOMOUS SEQUENTIAL INTER-NANO-UNIT CHEMICAL INTERACTIONS: FROM ONE NEUTRON CONSTRUCTION REMOVE ONE PROTON, AND ADD ONE NEUTRINO, WILL PRODUCE ONE ELECTRON. THIS PHENOMENON IS ELECTRON-GENESIS,

NEUTRON - PROTON + Electron Neutrino -----------→ ELECTRON

←-----------

CHEMISTRY OF NANO -UNITS

POSITRON – GENESIS

UNDER DEGENERATIVE AUTONOMOUS SEQUENTIAL INTER NANO –UNIT CHEMICAL INTERACTIONS CYCLES: FROM ONE PROTON CONSTRUCTION REMOVE ONE NEUTRON, ALSO REMOVE ONE ELECTRON NEUTRINO, WILL PRODUCE ONE POSITRON. THIS IS PHENOMENON OF POSITRON - GENESIS.

PROTON - NEUTRON - Electron Neutrino ---------------→ POSITRON

←------------------

THE FUNDAMENTAL PARTICLES (F.P.)

THE EXISTING LAWS AND ORDERS OF MATTER IN UNIVERSE

FUNDAMENTAL PARTICLES OR UNITS OF THE MATTER

FUNDAMENTAL PARTICLES ARE UNITS OF THE MATTER, THE FUNDAMENTAL PARTICLES CONSTRUCT EXISTING DIFFERENT ELECTRONS, NUCLEONS PARTICLE COMPOUND CONSTRUCTIONS OF EXISTING THINGS IN ALL OVER UNIVERSE. EACH GIVEN FUNDAMENTAL PARTICLE IN ANY GIVEN PLANET POSSESSES WELL- DEFINED, PARTICLE -SPECIFIC RELATIVELY FIXED PHYSICAL, CHEMICAL, AND BIOLOGICAL PROPERTIES, THE FUNDAMENTAL PARTICLES O DIFFERENT PLANETS ARE DIFFERENT FROM EACH OTHER, BUT EACH GIVEN PLANET POSSESSES PLANET- SPECIFIC, NUMEROUS DIFFERENT FUNDAMENTAL PARTICLES CLASSES, INDIGENOUS FOR THAT PLANET AT MOST.

THE FUNDAMENTAL PARTICLE'S WEIGHT, GRAVITY, FREQUENCY, WAVE-LENGTH, POLE, SPEED, SPIN, TRANSIT ROUTES, TILTS, ROTATIONS, COLOR, SIZE SHAPE, PARTICLE-COLONY SHAPE, PARTICLE COLONY SIZE, SENSIBILITY, PARTICLE CURRENTS, ETC., ARE ALL FUNDAMENTAL PARTICLE – SPECIFIC.

THE FUNDAMENTAL PARTICLES OF DIFFERENT PLANETS ARE DIFFERENT FROM EACH OTHER, THE FUNDAMENTAL PARTICLES AT DIFFERENT PLANETS CONSTRUCT DIFFERENT KIND PARTICLE COMPOUNDS, DIFFERENT ELECTRONS, NUCLEONS, AND ATOMS, UNDER AUTONOMOUS SEQUENTIAL INTER FUNDAMENTAL PARTICLE MOLECULAR CHEMICAL INTERACTION CYCLES AND CHAINS (A. S. I. F.P. Mol. C.I.C.). SOME PLANETS FUNDAMENTAL PARTICLES ARE BIOFRIENDLY AND SENSIBLE, MANY OTHERS PARTICLES ARE NOT SENSIBLE, MOST OTHER PLANETS PARTICLES ARE BIOHOSTILE, AND MANY ARE BIO-LETHAL PARTICLES.

FUNDAMENTAL PARTICLE COLONY.

PARTICLE POPULATION OF A COLONY.

FUNDAMENTAL PARTICLES TENDS TO COLONIZE IN WAVE –SHAPE FORMS, F.P. TRAVEL IN WAVE –SHAPE FORMS.

ONE FUNDAMENTAL PARTICLE COLONY EQUALS TO TOTAL FUNDAMENTAL PARTICLES POPULATION NUMBER IN ONE WAVE LENGTH FUNDAMENTAL PARTICLE, IN OTHER WORDS TOTAL PARTICLE NUMBERS THAT EXIST IN ONE WAVE – LENGTH PARTICLE IS EQUAL TO ONE COLONY FUNDAMENTAL-PARTICLE.

DIFFERENT FUNDAMENTAL PARTICLES HAVE DIFFERENT PARTICLE POPULATION NUMBERS PER COLONY AND DIFFERENT PARTICLE COLONIES, DIFFERENT WAVE –LENGTH SIZES SIMILAR TO DIFFERENT COLONY SIZES OF DIFFERENT KIND FUNDAMENTAL PARTICLES ARE DIFFERENT FROM EACH OTHER.

THE FUNDAMENTAL PARTICLE COLONY SIZE AND TOTAL FUNDAMENTAL PARTICLE-POPULATIONS NUMBERS AT EACH WAVE –LENGTH OF A GIVEN FUNDAMENTAL PARTICLE CLASS ALWAYS ARE WELL DEFINED, FIXED AND PARTICLE SPECIFIC. AGAIN COLONIES OF DIFFERENT PARTICLES ARE DIFFERENT FROM EACH OTHER.

EARTH'S FUNDAMENTAL PARTICLES.

PLANET EARTHS FUNDAMENTAL PARTICLES ARE LIGHT WEIGHT (LESS F.P. – MASS), WEAK, BIOFRIENDLY, LESS ENERGY CONTENT, LESS GRAVITON FORCE, SOME ARE FLOATING AIRBORN PARTICLES, SOME F.P. ARE SENSIBLE, BUT MOST EARTH'S FUNDAMENTAL PARTICLES ARE NON SENSIBLE AND ARE NOT DISCOVERED YET, ALMOST MAJORITIES OF EARTHS PARTICLES ARE BIOFRIENDLY, WITH EXCEPTIONAL BIOHOSTILE PARTICLE AS WELL, THESE ARE THE F.P. CONSTRUCT EARTHS ELECTRONS NUCLEONS PARTICLE COMPOUND CONSTRUCTIONS FOR BOTH LIVE AND NON-LIVE ELECTRONS NUCLEONS AND ATOMS.

FUNDAMENTAL PARTICLES WEIGHTS, GRAVITON, ENERGY CONTENTS ARE PARTICLES – SPECIFIC, FOR EXAMPLE THERMAL PARTICLES (T.-F.P.), LIGHT PARTICLES (Y.- F.P.), SONIC PARTICLES (S. –F.P.), ELECTRIC PARTICLES (E.-F.P.), X.-F.P., GAMMA FUNDAMENTAL PARTICLES ENERGIES, WEIGHTS, GRAVITON QUANTITIES AT EARTH ARE DIFFERENT FROM EACH OTHER, THESE PARTICLES IN COMPARISON TO BLACKS THERE IS THOUSANDS TO MILLIONS FOLDS DIFFERENCES, THOSE PARTICLES CONSTRUCT BLACK MATTER ATOMS FROM THEIR PARTICLE COMPOUNDS.

THE EARTH'S BIOHOSTILE FUNDAMENTAL PARTICLES SUCH AS GAMMA-PARTICLES, X- FUNDAMENTAL PARTICLES, INDUSTRIAL HIGH - ENERGY ELECTRIC FUNDAMENTAL PARTICLES, AND INDUSTRIAL HIGH ENERGY THERMAL - PARTICLES, ETC. AT EARTH MOSTLY ARE STORED INSIDE THE ELECTRONS AND NUCLEONS, IN PARTICLE COMPOUND COSTRUCTIONS FORMS.

PLANET EARTH'S BIOFRIENDLY SENSIBLE FUNDAMENTAL PARTICLES

INTRODUCTIONS TO VISIBLE BIOFRIENDLY LIGHT FUNDAMENTAL PARTICLES

AUDIBLE SONIC FUNDAMENTAL PARTICLES, AND THERMAL PARTICLES,

THERE ARE NUMEROUS DIFFERENT KIND FUNDAMENTAL PARTICLE CLASSIFICATIONS IN PLANET EARTHS, FOLLOWING ARE SOME EXAMPLES:

1 - CLASIFICATIONS OF LIGHT PARTICLES OVER SENSIBILITIES AND NON-SENSIBILITIES CHARACTERS.

2 – CLASSIFICATIONS OF VISIBLE LIGHT FUNDAMENTAL PARTICLES OVER EXISTING DIFFERENCES IN LIGHT FUNDAMENTAL PARTICLES COLORS, SUCH AS: RED COLOR FUNDAMENTAL PARTICLES CLASSES, YELLOW COLOR LIGHT FUNDAMENTAL PARTICLES CLASS, GREEN COLOR LIGHT PARTICLES CLASS, BLUE COLOR FUNDAMENTAL PARTICLES CLASS, VIOLET COLOR FUNDAMENTAL PARTICLES, AND WHITE COLOR LIGHT PARTICLES, ETC.

ABOVE DIFFERENT CLASSES PARTICLE WAVE LENGTH, PARTICLE FREQUENCY, PARTICLE ENERGY KIND AND ENERGY QUANTITY, PARTICLES GRAVITON FORCE, PARTICLE COLONY SIZES, PARTICLE COLONY SHAPES, ETC., ALL ARE DIFFERENT IN DIFFERENT COLOR PARTICLES FROM EACH OTHER.

3 - LIGHT PARTICLES CLASSIFICATIONS OVER BEING BIOHOSTILE OR BIOFRIENDLY IN EARTH.

4- SONIC FUNDAMENTAL PARTICLES CLASIFICATION, EITHER ARE HUMAN SENSIBLE OR NOT.

5- CLASSIFICATION OF SONIC FUNDAMENTAL PARTICLES OVER BEING BIOFRIENDLY OR BIOHOSTILE.

6- PLANET EARTHS THERMAL FUNDAMENTAL PARTICLE CLASSIFICATIONS.

7- PLANET EARTHS ELECTRIC FUNDAMENTAL PARTICLES CLASSIFICATIONS.

8- ETC.,

IN EARTH BIOFRIENDLY ELECTRIC FUNDAMENTAL PARTICLES CLASSES ARE IN CHARGE OF OPERATIONS OF ALL EARTHS LIVING THINGS BODY ORGANS AND BODY SYSTEMS OPERATIONS AS WELL AS ARE IN CHARGES OF CENTRAL AND PERIPHERAL INTELLIGENCE SYSTEMS CENTERS OPERATIONS, WITHOUT ELECTRIC BIOFRIENDLY FUNDAENTAL PARTICLES LIFE CAN NOT EXIST.

BIOFRIENDLY LIGHT FUNDAMENTAL PARTICLES, BIOFRIENDLY SONIC FUNDAMENTAL PARTICLES AND LIGHT-SONIC FUNDAMENTAL PARTICLE INFORMATION IMAGE PARTICLE CLOUDS (S.Y. – F.P. – I.I. – P.cl.) ARE IN CHARGE OF PRODUCTIONS OF THOUGHT CURRENTS AND PSYCHE IN LIVING THINGS CNS, AS WELL AS MOST OF CNS INTERNAL ELECTRONS NUCLEON PARTICLE COMPOUNDS CONSTRUCTIONS ARE CONSTRUCTED FROM S.Y.- F.P. – I.I. – P.cl. COMPOUNDS, WITHOUT SONIC PARTICLE CLOUDS AND WITHOUT LIGHT PARTICLE CLOUDS AND THERMAL-ELECTRIC PARTICLE CLOUDS PSYCHE AND THOUGHT CURRENTS COULD NOT DEVELOP AND CAN NOT EXIST AT ALL.

OTHER CLASSES FUNDAMENTAL PARTICLE IN EARTH

THERE ARE OTHER PARTICLE CLASSES, SOME ARE AIRBORN BIOFRIENDLY PARTICLES AND OTHERS ARE BIOHOSTILE PARTICLES CONSTRUCT DIFFERENT LIVE OR NON-LIVE ELECTRON-NUCLEON PARTICLE COMPOUND CONSTRUCTIONS IN EARTH.

1 – AIRBORN INFRA- RED FUNDAMENTAL - PARTICLES (Y. i. r. – F.P. = T. i. r. – F.P.) WHICH CAN TRAVEL AND TRANSMIT THROUGH OTHER MEDIA ROUTES AS WELL, THESE PATICLES CLASSIFIED UNDER LIGHT PARTICLES ALSO IN FACT ARE THERMAL FUNDAMENTAL PARTICLE AS WELL.

INFRA-RED FUNDAMENTAL PARTICLES ARE THERMAL ENERGY CONTENTS PARTICLES, INFRA RED CAN PROVIDE AND DELIVER QUANTUM THERMAL ENERGY FOR ANY GIVEN NANO-LOCATIONS INSIDE ELECTRONS – NUCLEONS WHO ARE IN NEEDS OF THERMAL – ENERGIES TO PERFORM NANO-TASKS, OR TRIGGERING START BIOLOGICAL CHEMICAL INTERACTIONS INSIDE ELECTRONS-NUCLEONS SUBSYSTEM –UNITS CHEMICAL LAB.S WHO ARE IN NEED OF QUANTUM DOSE TEMPERATURES TO INITIATE A. S. I. P. F. Mol. C.I.C., OR THROUGH PROVIDING QUANTUM HEAT ENERGIES FOR ANY GIVEN QUANTUM –LOCATIONS INFRA RED CAN TRIGGER START AND ACHIEVE DIFFERENT THERMO-DYNAMI1C PHYSICAL TASKS, OR DO OTHER THERMAL – ELECTRICAL MIXED TASKS (QUANTUM THERMO-ELECTRIC TASKS) INSIDE ATOMS, CELLS, ETC.

2 - FROM OTHER AIRBORN FUNDAMENTAL PARTICLE CLASSES OF PLANET EARTH ARE ULTRA-VIOLET FUNDAMENTAL PARTICLE (Y. u. v. – F.P. = E. u. v. – F.P.) CLASS, ULTRAVIOLET IS WEAK ELECTRIC FUNDAMENTAL PARTICLES, E. u. v. – F.P. BELONG TO ELECTRICAL FUNDAMENTAL PARTICLES CLASSES OF EARTH.

ULTRAVIOLET IS THE WEAKEST ELECTRIC FUNDAMENTAL PARTICLE IN PLANET EARTH, IT IS LOCATED THERE AFTER THE LIGHT FUNDAMENTAL PARTICLE CLASSES OF PLANET EARTH, ONE ULTRAVIOLETS FUNDAMENTAL PARTICLES ELECTRIC ENERGY CONTENT IS EQUAL TO ONE-UNIT ELECTRIC ENERGY, (THE QUANTITY OF ELECTRIC –ENERGY IN ONE ULTRAVIOLET PARTICLE IS EQUAL TO ONE UNIT ELECTRIC-ENERGY).

ULTRA VIOLET PARTICLES ARE ELECTRIC ENERGY PROVIDERS FOR QUANTUM SITES INSIDE ELECTRONS NUCLEONS AND CAN BE USED AS ELECTRIC ENERGY PROVIDER FOR INTERNAL ELECTRON NUCLEON NANO-LOCATIONS FOR ACHIEVING INTERNAL ELECTRON- NUCLEON NANO – WORKS OF DIFFERENT KINDS.

3 - MOST OF EARTHS PARTICLES EITHER BIOFRIENDLY OR BIOHOSTILE PRESENTLY HAVE NOT BEEN DISCOVERED AND NOT KNOWN YET.

THE NON DISCOVERED BIOFRIENDLY FUNDAMENTAL PARTICLE ARE BUILDING BLOCK OF ELECTRONS – NUCLEONS PARTICLE COMPOUND CONSTRUCTIONS AND ARE CREATORS WHO CONSTRUCTED MOST OF DIFFERENT LIVE ELECTRONS- NUCLEONS PARTICLE COMPOUND CONSTRUCTIONS.

EARTHS BIOFRIENDLY SENSIBLE SONIC FUNDAMENTAL PARTICLES AND OTHER CLASS PARTICLES

1 - PLANET EARTHS SONIC FUNDAMENTAL PARTICLES MOSTLY ARE LESS - WEIGHT LESS ENERGY AND LESS GRAVITON BIOFRIENDLY BENIGN WEAK SOUND FUNDAMENTAL PARTICLES (S. – F.P.), THE SONIC FUNDAMENTAL PARTICLES INFORMATION – IMAGE – PARTICLE –CLOUDS (S. –F.P. – I.I.- P.cl.) CONSTRUCT MOST OF INTERNAL ELECTRON-NUCULEONS CNS- PARTICLE COMPOUND MOLECULAR CONSTRUCTIONS AT AUDITARY CNS CENTERS ATOMS OF BRAIN CELLS.

2 - ADDITIONALLY, THE SONIC – LIGHT FUNDAMENTAL PARTICLE INFORMATION IMAGE PARTICLE CLOUDS (S. Y. – F.P. – I.I. – P. cl.) JOINTLY CONSTRUCT THOUGHT CURRENTS AND PSYCHE AT MOST PLANTS AND ANIMAL SPECIES, THERE ARE FEW OTHER BIOFRIENDLY EARTHS FUNDAMENTAL PARTICLES SUCH AS THERMAL FUNDAMENTAL PARTICLES (T. – F.P.) BIOFRIENDLY ELECTRIC FUNDAMENTAL PARTICLES (E. – F.P.) ALSO PARTICIPATE IN PSYCHE – GENESIS PHENOMENON, NEO- GENESIS OF INTERNAL ELECTRON- NUCLEON PARTICLE COMPOUNDS CONSTRUCTIONS THROUGH USE OF MOLECULAR S. Y. – F.P. – I.I. – P.cl. _ COMPOUND CONSTRUCTIONS ALSO ARE F.P. –I.I.- P.cl. STRORAGE PHENOMENON AS WELL.

3 – IN EARTHS ATMOSPHERE THERE IS NOT AIRBORN FREE HIGH ENERGY BIOHOSTILE SONIC FUNDAMENTAL PARTICLES, THIS WAS ALSO A FACTOR HELPED FOR SYNTHESIS OF LIVING THINGS IN EARTH, WEAK BIOFRIENDLY SONIC PARTICLE COMPOUNDS ARE USED IN CONSTRUCTION OF EARTHS ELECTRONS NUCLEONS.

4 - SUPERSONIC FUNDAMENTAL PARTICLES, MICROWAVE PARTICLES MOSTLY HAVE BEEN USED IN COMMERCIAL, MEDICAL OR NON MEDICAL INDUSTRIAL AND TELECOMMUNICATION FIELDS, THESE FUNDAMENTAL PARTICLES AS WELL AS THERE ARE LARGE OTHER UNKNOWN SONIC FUNDAMENTAL PARTICLE CLASSES IN EARTH WHICH ARE NON SENSIBLE TO MAN KIND SPECIES, STILL SOME OF THOSE SONIC PARTICLES ARE SENSIBLE TO ANOTHER ANIMALS.

5 - NON DISCOVERED FUNDAMENTAL PARTICLES NUMBERS IN EARTH FAR EXCEED FEW ABOVE MENTIONED KNOWN FUNDAMENTAL PARTICLE CLASSES, THERE ARE FAR LARGE NUMBERS OF OTHER BIOFRIENDLY FUNDAMENTAL PARTICLES ARE NON SENSIBLE AND NOT DISCOVERED BIOHOSTILE PARTICLES IN EARTH, AND ARE USED IN PARTICLE – COMPOUNDS CONSTRUCTING OF PLANET EARTHS DIFFERENT ATOMS ELECTRONS NUCLEONS PARTICLE COMPOUND CONSTRUCTIONS.

6 – DESCRIPTION OF LARGE NUMBERS OF DIFFERENT BIOHOSTILE FUNDAMENTAL PARTICLES CLASSES SUCH AS, X-PARTICLES CLASS, GAMMA PARTICLES CLASSES, LASER, INDUSTRIAL ELECTRIC PARTICLES, HARSH INDUSTRIAL THERMAL PARTICLES CLASSES HAS NO USE FOR THIS BOOK.

NANO-UNITS

=================

ELECTRONS, POSITRONS, PROTONS, NEUTRONS, QUARKS, NEUTRINOS, SUBSYSTEMS NANO – UNITS, LEPTONS, ANTI- ELECTRON NEUTRINOS, ELECTRON SYSTEMS, BARYONS, NUCLEON SYSTEMS AND SUBSYSTEMS NANO-UNITS, MESONS, ATOMS, ATOM-BASE MOLECULES, FUNDAMENTAL BASE MOLECULES, PLUS MANY OTHER KNOWN OR NON-KNOWN QUANTUM SIZE STRUCTURES ARE ALL EXAMPLES OF NANO-UNITS. PRESENTLY MOST NANO-UNITS ARE NO DISCOVERED AND NOT KNOWN YET, AND THE ABOVE ARE SOME EXAMPLES OF NANO-UNITS.

ALL NANO-UNIT ARE INDEPENDENT QUANTUM SIZE STRUCTURES. FUNDAMENTAL PARTICLE CONSTRUCTIONS OF NANO –UNIT ALL ARE WELL- DEFINED FIXED NANO-UNIT SPECIFIC PARTICLE –COMPOUND CONSTRUCTIONS. ALL NANO-UNITS POSSESS NANO-UNIT SPECIFIC PHYSICAL CHEMICAL BIOLOGICAL PROPERTIES.

CONSTRUCTION OF NANO –UNITS

MOSTLY DIFFERENT KIND FUNDAMENTAL PARTICLES SUCH AS ELECTRIC PARTICLE, LIGHT PARTICLES, SONIC PARTICLES, THERMAL PARTICLES, ETC. JOINTLY CONSTRUCT DIFFERENT NANO-UNITS, ELECTRONS. NUCLEONS PARTICLE COMPOUND CONSTRUCTIONS, STILL THERE ARE MANY ELECTRONS, NUCLEONS, AND ATOMS WHOSE PARTICLE COMPOUNDS HAVE BEEN CONSTRUCTED FROM FEW SEVERAL DIFFERENT KIND FUNDAMENTAL PARTICLE COMPOUND CONSTRUCTED ELECTRON NUCLEON CONSTRUCTIONS, SUCH AS ELECTRIC PARTICLE ATOMS, ELECTRONS, NUCLEONS, OR THERMAL PARTICLE COMPOUND ELECTRONS AND NUCLEONS, OR SONIC PARTICLE ATOMS, AND LIGHT PARTICLE COMPOUND CONSTRUCTED ELECTRONS NUCLEONS AND ATOMS, ETC.

ELECTRONS NUCLEONS FUNDAMENTAL PARTICLE CONSTRUCTIONS

THE FUNDAMENTAL PARTICLES CONSTRUCTIONS OF ELECTRONS, NUCLEONS, ATOMS, NANO-UNITS

THE SPECIFIC FUNDAMENTAL PARTICLE CONSTRUCTED ELECTRONS, NUCLEONS, ATOMS, NANO-UNITS

MOST OF EARTH'S ELECTRONS NUCLEONS PARTICLE COMPOUNDS CONSTRUCTED FROM NURMEROUS DIFFERENT KIND BIOFIENDLY FUNDAMENTAL PARTICLES, THE FOLLOWING FUNDAMENTAL PARTICLE COMPOUNDS SUCH AS: THE ELECTRIC PARTICLE COMPOUNDS, THE LIGHT PARTICLE COMPOUNDS, SONIC PARTICLE COMPOUNDS, THE THERMAL PARTICLE COMPOUNDS, OCCASIONALLY X-PARTICLES AND GAMMA PARTICLES COMPOUNDS, ETC., ARE THE MOST COMMONLY USED PARTICLES COMPOUNDS THAT CONSTRUCTING THE EARTHS ELECTRONS NUCLEONS PARTICLE COMPOUND AT EARTH, CONTRARY TO OTHER PLANETS WHICH THEIR PARTICLE CONSTRUCTED MOLECULAR COMPOUNDS MOSTLY ARE BIOHOSTILE.

ALSO THERE ARE MANY ELECTRONS NUCLEONS, THAT THEIR FUNDAMENTAL PARTICLE COMPOUNDS MOLECULAR COSTRUCTIONS ARE MADE FROM A FEW KNOWN SPECIFIC FUNDAMENTAL PARTICLES, SUCH AS H-O-H, OR CRYSTALINE'S ELECTRONS NUCLEONS, WHICH THE H- O –H AND CRYSTALINES DOMINENT CONSTRUCTING FUNDAMENTAL PARTICLES ONLY ARE DIFFERENT CLASS LIGHT FUNDAMENTAL PARTICES, THESE ATOMS ELECTRONS NUCLEONS HAVE BEEN CONSTRUCTED FROM MANY DIFFERENT KIND LIGHT PARTICLE COMPOUNDS, SUCH AS RED

COLOR LIGHT PARTICLE COMPOUNDS, BLUE OR VIOLET PARTICLE COMPOUNDS, YELLOW OR GREEN PARTICLE COMPOUNDS, ETC.

WHEN THE H- O –H, OR CRYSTALINE ELECTRONS NUCLEONS LIGHT- PARTICLE _ COMPOUNDS COMBINE WITH DIFFERENT TYPES OTHER AIRBORN LIGHT –PARTICLES, X - FUNDAMENTAL PARTICLES, GAMMA PARTICLES, OR ELECTRIC PARTICLES, ETC., THROUGH A. S. I. F.P. Mol. C.I.C., THE CHEMICAL COMBINATIONS PARTICLES BY PRODUCTS ARE RELEASE OF DIFFERENT KIND AND COLOR FREED LIGHT PARTICLES, SENSIBLE BY NAKED EYES.

IN EARTH, NUMEROUS ATOMS, ELECTRONS, NUCLEONS, PARTICLE COMPOUNDS ARE CONSTRUCTED FROM BIOFRIENDLY DIFFERENT TYPE PARTICLES, SUCH AS THERMAL PARTICLES, OR BIOFRIENDLY ELECTRIC PARTICLES, SONIC PARTICLES, LIGHT PARTICLES, ETC. ALSO SOME OTHER ELECTRONS NUCLEONS PARTICLE COMPOUND CONSTRUCTED FROM BIOHOSTILE PARTICLES, SUCH AS X-PARTICLE COMPOUNDS, GAMMA PARTICLE COMPOUNDS, HARSH THERMAL OR ELECTRICAL PARTICLE COMPOUND CONSTRUCTION ELECTRONS, NUCLEONS, AND ATOMS.

LIGHT FUNDAMENTAL PARTICLE CONSTRUCTED ATOMS, ELECTRONS, NUCLEONS, NANO-UNITS

LIGHT PARTICLE NANO-UNITS, LIGHT ENERGY PROVIDER ATOMS

CRYSTALINE ATOMS ARE ONE EXAMPLE FROM LIGHT-FUNDAMENTAL PARTICLE CONSTRUCTED ATOMS, CRYSTALINE ATOMS INTERNAL ELECTRONS -NUCLEONS SUBSYSTEM –UNITS PARTICLE- COMPOUNDS MOSTLY ARE BUILT FROM SUN ORIGIN LIGHT FUNDAMENTAL PARTICLE-COMPOUNDS, CRYSTALINE ELECTRONS-NUCLEONS MOSTLY CONSTRUCTED FROM RED, YELLOW, GREEN, BLUE, VIOLET, ETC. COLOR LIGHT PARTICLES – COMPOUNDS AS WELL AS MANY OTHER NON-DISCOVERED NON-SENSIBLE SUN-ORIGIN PARTICLE- COMPOUNDS WHICH THOSE ALSO ARE THE MAIN FUNDAMENTAL PARTICLES WHICH PARTICIPATED AT CONSTRUCTION OF THE CRYSTALNE ATOMS PARTICLE STRUCTURES.

THE LIGHT FUNDAMENTAL- PARTICLES ARE NEEDED FOR BUILDING THE PLANTS ELECTRONS – NUCLEONS LIGHT PARTICLE COMPOUND CONSTRUCTIONS, IN ANIMAL SPECIES LIGHT FUNDAMENTAL PARTICLE PRODUCE INFORMATION IMAGE PARTICLE CLOUDS, WHICH THE LIGHT INFORMATION IMAGE PARTICLE –CLOUDS ARE MAIN ELEMENTS OF THOUGHT CURRENTS AS WELL AS THE LIGHT PARTICLE CLOUDS ALSO ARE USED IN CONSTRUCTIONS OF THE CNS ELECTRONS – NUCLEONS PARTICLE COMPOUND CONSTRUCTIONS, ALSO LIGHT FUNDAMENTAL PARTICLES INFORMATION IMAGE PARTICLE CLOUDS CURRENTS CONSTRUCT PSYCHE AND THOUGHT CURRENTS.

ELECTRIC FUNDAMENTAL PARTICLE CONSTRUCTED ELECTRONS, NUCLEONS, ATOMS, NANO-UNITS

ELECTRIC ENERGY PROVIDER ATOMS, AND THE NANO BATTERIES

THE ATOMS CONSTRUCTED FROM ELECTRIC FUNDAMENTAL PARTICLES ARE THE ELECTRIC PARTILCE ATOMS, THES ATOMS ELECTRONS NUCLEONS PARTICLE COMPOUND CONSTRUCTIONS ARE MOSTLY BUILT FROM DIFFERENT CLASSES OF ELECTRIC FUNDAMENTAL PARTICLE, AND THE ELECTRIC FUNDAMENTAL PARTICLES OF THESE ELECTRI MOLECULAR COMPOUND CONSTRUCTIONS INHERITANTLY POSSESSES QUANTUM ELECTRIC ENERGY CONTENS AT THE PARTICLES, THESE ARE QUANTUM QUANTITY ELECTRIC ENERGY SOURCES INSIDE ELECTRONS, NUCLEONS, NANO-UNIT, AND CAN GENERATE ELECTRIC ENERGY FOR ANY GIVEN ELECTRONS, NUCLEONS, NANO – SITES.

THROUGH DEGENERATIVE A. S. I. F.P. Mol. C. I. C. THESE ELECTRIC FUNDAMENTAL-PARTICLES RELEASE FROM PARTICLE-COMPOUND CONSTRUCTIONS FORMS, INTO FREE ELECTRIC FUNDAMENTAL-PARTICLE FORMS, RELEASE ELECTRIC FUNDAMENTAL ARTICLES TRAVEL IN PARTICLE CURRENT CIRCULATION SYSTEMS, AND CAN BE USED AT INTER-ELECTRON INTER-NUCLEON CHEMICAL LAB.S AS ENERGY SOURCES OR CAN BE USED AS QUANTUM-ENERGY SOURCES AND ELECTRIC ENERGY PROVIDER IN INTERNAL-ATOM ELECTRIC BATTERIES OR USED IN CONSTRUCTION OF INTERNAL-ATOM INDUSTRIAL NANO-BATTERIES AND IN CONSTRUCTION OF ANY OTHER NANO- STRUCTURES.

ELECTRIC FUNDAMENTAL PARTICLES HAS UTMOST IMPORTANCE FOR ANIMAL ORGAN FUNCTIONS, ALSO ELECTRI PARTICLE PROVIDER ATOMS WHICH THEIR PARTICLE CONSTRUCTIONS HAVE BEEN BUILT BY DIFFERENT ELECTRIC PARTICLE COMPOUNDS, ARE ALL CRITICALLY NEEDED FOR MAINTAINING AUTONOMOUS FUNCTIONS OF THE DIFFERENT BODY ORGAN AND SYSTEM, THEY PROVIDE ELECTRIC ENERGY FOR DIFFERENT QUANTUM –LOCATIONS AND CONDUCT PHYSICAL, CHEMICAL, BIOLOGICAL NANO- FUNCTIONS, AT DIFFERENT QUANTUM LOCATIONS.

THE PLANTS, ANIMAL SPECIES WITHOUT BIOFRIENDLY ELECTRIC PARTICLES AND CURRENTS, WHICH ARE PRODUCERS OF NANO-NEURAL CURRENTS AS WELL OERATERS OF NANO-ELECTRICAL TASKS FOR DIFFERENT BODY ORGANS AND SYSTEMS IN QUATUM LOCATIONS, CAN NOT EXIST, AND CAN NOT SURVIVE EVEN FOR A SECOND, WITHOUT ELECTRIC PARTICLES EXISTENCES.

IN INDUSTRIAL WORLD, THE ELECTRONS, NUCLEONS, ATOMS, AND ELECTRIC FUNDAMENTAL PARTICLES, WHICH THESE NANO-UNITS ELECTRIC PARTICLE COMPOUNDS HAVE BEEN CONSTRUCTED FROM HIGH ENERGY POWERFU HIGH WEIGHT, HIGH GRAVITON ELECTRIC PARTICLE CONSTRUCTIONS, THESE HIGH ELECTRIC ENERGY PARTICLES AND ELECTRIC PARTICLE COMPOUNDS HAVE BEEN USED TO PERFORM DIFFERENT KIND HIGH ELECTRIC ENERGY INDUSTRIAL ELECTRO-MECHANIC, THERMO-ELECTRIC, COMMERCIAL, INDUSTRIAL TASKS, AS WELL IN ELECTRIC GENERATOR PLANTS.

BY NO WAY THE ELECTRIC PARTICLES ENERGY AT EARTH ARE COMPARABLE TO OTHER VICIOUS HIGH ELECTRIC-ENERGY CONTENTS FUNDAMENTAL PARTICLES, AT HARSH PLANETS IN DISTANT GALAXIES NEXT TO BLACK HOLES THE ELECTRICAL ENERGY CONTNTS OF ELECTRIC FUNDAMENTAL PARTICLES ARE MILLIONS FOLDS STRONGER THA EARTHS BIOFREINDLY WEAK ELECTRIC FUNDAMENTAL PARTICLES ELECTRIC ENERGY.

THE THERMAL FUNDAMENTAL PARTICLE CONSTRUCTED ELECTRONS, NUCLEONS, ATOMS, NANO-UNITS.

THERMAL ENERGY PROVIDERS TO NANO-LOCATIONS (QUANTOM SITES

THE BIOFREINDLY WEAK THERMAL PARTICLES SUCH AS, INFRA - RED THERMAL FUNDAMENTAL PARTICLES, OR OTHER HARSH BIOHOSTILE THERMAL PARTICLES INTER INTO INTERNAL ELECTRONS NUCLEONS SUBSYSTEMS UNIT CHEMICAL LAB.S, AND COMBINE WITH INTERNAL ELECTRON NUCLEON PARTICLE COMPOUNDS, PRODUCE INFRA-

RED PARTILE COMPOUNDS, OR ANY OTHER THERMAL PARTICLE COMPOUNDS INSIDE ELECTRONS AND NUCLEONS THROUGH A. S. I. F.P. Mol. C.I.C.

THE INFRA-RED PARTICLES COMPOUNDS INSIDE ELECTRONS NUCLEONS ARE STORAGE OF HEAT PROVIDER PARTICLES INSIDE ELECTRONS NUCLEONS IN THERMAL PARTICLE COMPOUND FORMS, THESE THERMAL PARTICLE COMPOUNDS ARE THERMAL ENERGY PROVIDERS AND LIGHT ENERGY PROVIDER PARTICLES, READY TO DELIVER HEAT INTO ANY GIVEN QUANTUM LOCATIONS TO ACHIEVE NEEDED NANO-TASKS.

THE THERMAL PARTICLE COMPOUND CONSTRUCTED NANO-UNITS, ATOMS, ELECTRONS, NUCLEONS AT BIOLOGICAL WORLD ARE USED AS PROVIDERS AND GENERATORS OF THE QUATUM THERMAL-ENERGY SOURCES FOR PERFORMING INNORMOUS DIFFERENT KINDS BIOLOGICAL, CHEMICAL AND PHYSIOLOGICAL FUNCTIONS, UNDER DEGENERATIVE A. S. I. F.P. Mol. C.I.C. THE THERMAL PARTICLES RELEASE AS FREE HEAT PROVIDING FUNDAMENTAL PARTICLES, AND CAN PROVIDE THERMAL-ENERGY FOR ANY NANO-- LOCATION AT ANY GIVEN INTER ELECTRON NUCLEON QUANTUM LOCATIONS, TO DELIVER ON SITE HEAT –ENERGY FOR CHEMICAL INTERACTIONS, OR OTHER QUANTUM THERMO-DYNAMIC BIOLOGICAL – FUNCTIONS.

THE X- FUNDAMENTAL PARTICLE CONSTRUCTED ATOMS, GAMMA FUNDAMENTAL PARTICLE CONSTRUCTED ATOMS, AND OTHER HARSH BIOCIDAL CONSTRUCTED ATOMS ARE NOT DESCRIBED.

QUANTUM – ENERGY PROVIDERS, QUANTUM – ENERGY RECIPIENTS OF NANO – SITES

FOR PERFORMING NEEDED NANO-TASKS AT NANO – LOCATIONS

NANO-UNIT'S ENERGY SOURCES QUANTUM ENERGY PROVIDERS TO: PERIPHERON, NUCLEUS, ELECTRONS AND NUCLEONS

PERFORMING PHYSICAL, CHEMICAL BIOLOGICAL TASKS, INSIDE ELECTRONS –NUCLEONS LAB.S

UNDER PROVIDED NANO – ENERGIES BY NANO- ENERGY PROVIDERS

THERMAL FUNDAMENTAL PARTICLES STORE HEAT –ENERGY IN FORMS OF THERMAL PARTICLE COMPOUNDS INSIDE ELECTRONS- NUCLEONS AND PROVIDE THERMAL ENERGIES FOR NANO –LOCATIONS WHEN IT IS NEEDED, ELECTRIC PARTICLES PROVIDE ELECTRICITY ENERGY IN QUANTUM AMOUNTS TO BE USED INSIDE DIFFERENT QUANTUM LOCATIONS FOR ACHIEVING DIFFERENT KIND NANO-TASKS INSIDE DIFFERENT NANO-UNITS, LIGHT FUNDAMENTAL PARTICLES SONIC PARTICLES AND ALL OTHER FUNDAMENTAL PARTICLES DO THE SAME.

UNIT OF THERMAL – ENERGY

INFRA- RED THERMAL FUNDAMENTAL PARTICLES

THE THERMAL ENERGY CONTENT OF ONE INFRA- RED PARTICLE IS EQUALS TO ONE THERMAL –ENERGY-UNIT

THE EXISTING THERMAL ENERGY AMOUNT IN ONE INFRA- RED THERMAL FUNDAMENTAL PARTICLE IS EQUAL TO ONE UNIT THERMAL ENERGY.

INFRA – RED FUNDAMENTAL PARTICLE (T. i. r. – F.P.) IS A BIOFRIENDLY WEAK THERMAL FUNDAMENTAL PARTICLE IN EARTH, THE INFRA-RED SIMILAR TO OTHER THERMAL PARTICLES PROVIDES QUANTUM THERMAL ENERGY INTO INTERNAL ELECTRON NUCLEON QUANTUM LOCATIONS (NANO-SITES) TO BE USED TO PERFORM DIFFERENT BIOLOGICAL CHEMICAL AND PHYSICAL NANO-TASKS, THROUGH THE USE OF PARTICLE PROVIDED THERMAL ENERGIES.

INFRA RED FUNDAMENTAL PARTICLE COMPOUNDS (T. i. r. – F. P. _ COMP.) SIMILAR TO OTHER PARTICLES HAVE BEEN USED IN CONSTRUCTIONS OF DIFFERENT INTERNAL ELECTRONS-NUCLEONS PARTICLE COMPOUNDS STRUCTURES (THIS IS PHENOMENON OF THERMAL –ENERGY STORAGE PHENOMENON INSIDE PARTICLE COMPOUNDS) AND UNDER DEGENERATIVE A. S. I. F.P. Mol. C.I.C. THESE PARTICLE COMPOUNDS CAN RELEASE INFRA RED PARTICLES FREE IN INTERNAL ELECTRON-NUCLEON NANO-SITES, AS THERMAL ENERGY PROVIDER IN ANY GIVEN HEAT ENERGY NEEDED NANO-LOCATIONS.

BIOFRIENDLY RED COLOR FUNDAMENTAL PARTICLE

RED COLOR LIGHT FUNDAMENTAL PARTICLE (Y. r. –F.P.) OR RED COLOR THERMAL FUNDAMENTAL PARTICLE (T. r. – F.P.), THE RED COLOR FUNDAMENTAL PARTICLES POSSESSES TWO DIFFERENT KIND BIOFRIENDLY THERMAL ENERGY, AND BIOFRIENDLY LIGHT ENERGIES.

THE INTERNAL ELECTRON'S - NUCLEON'S RED LIGHT PARTICLE COMPOUNDS ARE QUANTUM THERMAL ENERGY STORAGE SYSTEMS, AND QUANTUM ENERGY PROVIDERS BECAUSE OF THEIR INHERITANT ENERGY CONTENTS.

BIOFRIENDLY ELECTRIC ENERGY PROVIDER PARTICLES TO NANO-SITES

INTERNAL ELECTRON NUCLEON ENERGY STORAGE SYSTEMS AND ENERGY PROVIDER SYSTEMS

BLUE COLOR PARTICLES AND VIOLET COLOR PARTICLES ARE ELECTRIC ENERGY DONNOR PARTICLES

DIFFERENT FUNDAMENTAL PARTICLES POSSESSES DIFFERENT KIND ENERGY QUALITIES AND QUANTITIES, FOR EXAMPLE THE THERMAL ENERGY CONTENTS OF DIFFERENT FUNDAMENTAL PARTICLES AT DIFFERENT CLASSES THERMAL FUNDAMENTAL PARTICLES VARY REMARKABLY FROM EACH OTHER, WHEN COMPARING ONE PARTICLE CLASS FUNDAMENTAL PARTICLES ENERGY CONTENT WITH ANOTHER CLASS PARTICLES ENERGIES, THE RESULTS ARE DIFFERENT BUT ALWAYS THE FINDINGS ARE FUNDAMENTAL PARTICLE SPECIFIC.

SOME FUNDAMENTAL PARTICLES ARE MULTI – ENERGY PROVIDERS AND POSSESSES DIFFERENT KIND ENERGIES CONTENTS, THE BIOFRIENDLY RED COLOR FUNDAMENTAL PARTICLES POSSESSES TWO KIND THERMAL ENERGY AND LIGHT ENERGY BOTH, THE RED COLOR FUNDAMENTAL PARTICLES ARE THERMAL ENERGY PROVIDERS AS WELL AS LIGHT –ENERGY DONNOR'S, THE RED COLOR PARTICLES COMPOUNDS ALSO ARE THERMAL-ENERGY AND LIGHT ENERGY STORAGE SYSTEMS INSIDE THE ELECTRONS NUCLEONS PARICLE COMPOUND CONSTRUCTIONS.

THE VIOLET COLOR FUNDAMENTAL PARTICLES AND THE BLUE COLOR FUNDAMENTAL PARTICLES FROM PLANET SUN ORIGIN, BOTH ALSO ARE MULTI – ENERGY TWO DIFFERENT KIND FUNDAMENTAL PARTICLES AS WELL, BLUE COLOR FUNDAMENTAL PARTICLES AND VIOLET COLOR FUNDAMENTAL PARTICLES BOTH POSSESSES ELECTRIC- ENERGIES AS WELL AS LIGHT ENERGIES IN PARTICLE CONTENT, BLUE PARTICLES AND VIOLET PARTICLES ARE MULTI - ENERGY CONTENT PARTICLES FROM TWO DIFFERENT KIND PARTICLES CLASSES.

BOTH OF BLUE COLOR PARTICLES AND VIOLET COLOR PARTICLES ARE ELECTRIC- ENERGY PROVIDERS PARTICLES AS WELL AS LIGHT ENERGY DONNOR'S PARTICLES, BOTH BLUE PARTICLES AND VIOLET PARTICLES CAN GET INTO DIFFERENT NANO-LOCATIONS OF DIFFERENT ELECTRONS NUCLEONS AND PROVIDE NEEDED ELECTRIC ENERGIES OR LIGHT ENERGIES, TO ANY GIVEN NANO—LOCATIONS WHO ARE IN NEED OF ELECTRIC OR LIGHT ENERGIES TO DO QUANTUM NANO TASKS THROUGH USE OF THOSE PROVIDED ELECTRIC -LIGHT ENERGIES.

ENERGY SOURCES AND PROVIDERS FOR NANO TASKS

INSIDE ELECTRONS NUCLEONS, THE PARTICLES AT QUANTUM LOCATIONS

FUNDAMENTAL PARTICLES CONTROL CHEMICAL, PHYSICAL, BIOLOGICAL, TASKS THROUGH PROVIDING, OR STOPPING ENERGY SOURCES, TO DIFFERENT NANO- SITES FOR NANO- FUNCTIONS, INCLUDING A. S. I. F.P. Mol. C.I.C.

PHENOMENON OF ENERGY STORAGE- PARTICLES

PHENOMENON OF ENERGY STORAGE INSIDE ELECTRONS NUCLEONS, THE PARTICLES

FUNDAMENTAL PARTICLES NATURALLY POSSESS INHERITANT FUNDAMENTAL PARTICLE SPECIFIC ENERGIES, WEIGHT AND GRAVITON. FUNDAMENTAL PARTICLES TRANSIT THROUGH DIFFERENT PARTICLE TRANSMISSION ROUTES INTO INSIDE ELECTRONS NUCLEONS, AND UNDER REGENERATIVE A. S. I. F.P. Mol. C.I.C. COMBINE WITH INTER ELECTRON NUCLEON PARTICLE COMPOUNDS AND CONSTRUCT ELECTRONS NUCLEONS PARTICLE-COMPOUND CONSTRUCTIONS, THEREFORE FUNDAMENTAL PARTICLES STORE THEIR ENERGY CONTENT IN PARTICLE COMPOUND FORMS INSIDE ELECTRONS NUCLEONS, THIS IS PHENOMENON OF STORAGE OF ENERGY INSIDE ELECTRONS NUCLEONS.

PHENOMENON OF ENERGY RELEASE FROM NANO-UNITS TO DIFFERENT NANO-SITES BY PARTICLES

PHENOMENON OF ENERGY RETRIEVAL FROM ELECTRONS-NUCLEONS TO OUTSIDE

FUNDAMENTAL PARTICLES PRESENCE IN ANY GIVEN QUANTUM LOCATIONS INSIDE ELECTRONS NUCLEONS IS EQUAL TO EXISTENCES OF FUNDAMENTAL PARTICLE ENERGY IN THAT GIVEN NANO-SITE AS WELL, FUNDAMENTAL PARTCLES IN ANY GIVEN NANO-LOCATIONS CAN PROVIDE NEEDED ENERGY AND INITIATE QUANTUN PHYSICAL CHEMICAL OR BIOLOGICAL FUNCTIONS THROUGH THE USE OF EXISTING PARTICLE ENERGY AND PROVIDING

PARTICLE ENERIES TO ACCOMPLISH NEEDED CHEMICAL BIOLOGICAL OR PHYSICAL NANO-TASK ACCORDING THE PARTICLE MOLECULAR EVOLUTION ORDERS AND PATHS.

UNDER DEGENERATIVE A. S. I. F.P. Mol. C.I.C. THE CONSTRUCTED FUNDAMENTAL PARTICLES – COMPOUNDS BREAK DOWN INTO SMALLER FUNDAMENTAL PARTICLES FORMS AND RELEASE PARTICLE ENERGY INTO NANO-LOCATION, FUNDAMENTAL PARTICLES POSSESS ENERGY AND PROVIDE ENERGY INTO RELEASED QUANTUM LOCATIONS AS ENERGY DONNOR'S PROVIDE THEIR ENERGIES AND THROUGH USE OF THAT ENERGY ACHIEVE CHEMICAL PHYSICAL OR ANY OTHER NEEDED TASKS IN QUANTUM LOCATIONS THROUGH USE OF RELEASED PARTICLES ENERGIES.

IN OTHER SITUATIONS, WHEN RELEASED AND FREED FUNDAMENTAL PARTICLES LEAVE NANO SITES AND SCAPE TO OTHER LOCATIONS FROM NANO-SITES AS FREE PARTICLES, IN CASE OF ABSENCE OF FUNDAMENTAL PARTICLES TO ANOTHER LOCATIONS BY ITSELF CREATE ENERGY VACUUM AND NO ENERGY IN NANO-LOCATIONS, THEREFORE THE PHYSICAL CHEMICAL BIOLOGICAL FUNCTION ALL STOP THROUGH SOURCE CUT, IN THE ABCENSE OF FUNDAMENTAL PARTICLES NANO-TASKS STOP ACCORDINGLY.

THIS PHENOMENON SIMILAR TO ON-OFF KEY, CONTROL AUTONOMOUS SEQUENTIAL INTERACTIONS, START OR STOP NANO TASKS IN NANO-LOCATIONS, THROUGH PROVIDING FUNDAMENTAL PARTICLE ENERGY INTO GIVEN QUANTUM LOCATION TRIGGER START NANO-FUNCTIONS AUTONOMOUSLY, IN THE ABSENCE OF DONOR FUNDAMENTAL PARTICLE ENERGY THE AUTONOMOUS SEQUENCES TRIGGER STOP- FUNCTIONS IN NANO-LOCATIONS, IN START PROVIDE ENERGY AND IN STOP INTERRUPT ENERGY SOURCES.

RED COLOR FUNDAMENTAL PARTICLE'S OPERATED THERMAL – FUNCTIONS

UNDER REGENERATIVE A. S. I. F.P. Mol. C.I.C. THE RED COLOR FUNDAMENTAL PARTICLES INTER INTO CELLS ELECTRONS-NUCLEONS SUBSYSTEM UNITS COMBINE WITH PARTICLE COMPOUNDS AND PRODUCE RED COLOR PARTICLE COMPOUND CONSTRUCTIONS, AS RED COLOR PARTICLES ENERGY MOSTLY IS THERMAL ENERGY AS WELL AS LIGHT ENERGY, IN RESULT RED- PARTICLE COMPOUND CONSTRUCTION IS THERMAL –ENERGY STORAGE SYSTEM AS WELL, THIS PHENOMENON IS THERMAL –ENERGY STORAGE AND LIGHT-ENERGY STORAGE SYSTEMS PHENOMENON.

RED- PARTICLE- COMPOUND CONSTRUCTIONS OF ELECTRONS NUCLEONS ARE THERMAL – LIGHT ENERGIES STORAGE SYSTEMS, WHICH THROUGH DEGENERATIVE AUTONOMOUS SEQUENTIAL CHEMICAL INTERACTIONS CYCLES (DEG. A. S. I. F.P. Mol. C.I.C.) RED THERMAL PARTICLES CAN BE RELEASED IN ANY INTERNAL ELECTRON-NUCLEON NANO SITE PROVIDE ENERGY IN EXACT NEEDED QUANTUM-LOCATIONS OR THE RELEASED PARTICLES CAN INTERACT WITH OTHER PARTICLES AND PERFORM INNORMOUS DIFFERENT KINDS PHYSICAL CHEMICAL BIOLOGICAL FUNCTIONS, ALSO RELEASED PARTICLES PARTICIPATE IN INNORMOUS DIFFERENT KIND A. S.I. F.P. Mol. C.I.C..

THE STORED RED-COLOR PARTICLES WHEN RELEASED IN NANO-SITE PROVIDE THERMAL –ENERGY FOR LIVING THINGS INTERNAL ELECTRON – NUCLEON INTERNAL SUBSYSTEM-UNITS CHEMICAL LAB.S TO TRIGGER START AND CONDUCT A. S. I. F.P. Mol. C.I.C., ADDITIONALLY WHEN RED COLOR PARTICLE ALL CONSUMED AND NOT PRESENT IN QUANTUM LOCATIONS ANY MORE AND NO MORE THERMAL ENERGIES PRESENT, UNDER THESE CONDITIONS THE A.S. I. F.P. Mol. C.I.C. STOPS IMMEDIATELY AS WELL, THE RED COLOR THERMAL PARTICLES OPERATED BIOLOGICAL PHYSICAL CHEMICAL NANO-TASKS THROUGH PROVIDING THERMAL ENRGIES AND LIGHT ENEGIES AS WELL.

INFRA- RED AND RED COLOR FUNDAMENTAL PARTICLES AS WELL AS NUMEROUS OTHER NON- DISCOVERED THERMAL PARTICLES ALL TOGETHER WITH COOPERATIONS OF EACH OTHER PROVIDE QUANTUM THERMAL ENERGY FOR PERFORMING BIOLOGICAL CHEMICAL PHYSICAL NANO-TASKS AT INTERNAL ELECTRON- NUCLEON NANO – LOCATIONS OF LIVING THINGS CELLS.

ALL OTHER FUNDAMENTAL PARTICLES SUCH AS ELECTRIC PARTICLES, SONIC PARTICLES, X-PARTICLES, GAMMA PARTICLES, LIGHT PARTICLES, ETC., ALL FOLLOW THE SAME EXACT FUNDAMENTAL PARTICLE RULES IN OPERATIONS OF PHYSICAL CHEMICAL BIOLOGICAL TASKS IN INTERNAL ELECTRON NUCLEON NANO LOCATIONS SIMILAR TO RED COLOR FUNDAMENTAL PARTICLES AS EXPLAINED IN ABOVE.

PROVIDING NEEDED ANY KIND ENERGIES SUCH AS ELECTRIC, LIGHT, SONIC, X, GAMMA ENERGIES, ETC., IMMEDIATELY TRIGGER START DIFFERENT PHYSICAL CHEMICAL BIOLOGICAL NANO-TASKS IN GIVEN INTER

ELECTRON NUCLEON NANO-LOCATIONS AND ACCOMPLISH INTENDED QUANTUM TASK AND AT THE END, THE FUNDAMENTAL PARTICLES WHICH WAS PROVIDING DIFFERENT ELECTRIC-LIGHT- SONIC- GAMMA- X- ENERGIES TERMINATE AND TRIGGER STOP DIFFERENT ONGOING PHYSICAL CHEMICAL BIOLOGICAL NANO FUNCTIONS THROUGH LEAVING SCAPING AREA OR WHEN CONSUMED ALL IN DIFFERENT CHEMICAL PHYSICAL BIOLOGICAL INTERACTIONS THE END OF ENERGY SOURCE IS EQUAL END OF FUNCTIONS, IMMEDIATELY TERMINATE THE CONTINUATIONS OF THE ONGOING TASKS.

ALL FUNDAMENTAL PARTICLES PHYSICAL CHEMICAL BIOLOGICAL INTERACTIONS TAKE PLACE WITH EXPONENTIAL SPEED AUTONOMOUSLY AND SEQUENTIALLY, ALL INTERACTIONS EITHER IN LIVE OR NON-LIVE NANO-UNITS ALL FOLLOW THE SAME FUNDAMENTAL PARTICLE RULES AND LAWS ACCROSS THE UNIVERSE.

GREEN COLOR LIGHT FUNDAMENTAL PARTICLES (YG- FP)

QUANTUM LIGHT ENERGY POVIDER FUNDAMENTAL PARTICLES

UNIT OF LIGHT ENERGY

DEFINITION: THE EXISTING LIGHT ENERGY AMOUNT IN ONE GREEN LIGHT FUNDAMENTAL PARTICLE IS EQUAL TO ONE UNIT LIGHT- ENERGY.

GREEN LIGHT FUNDAMENTAL PARTICLE (Y. g. – F.P.) IS AN ESSENTIAL BIOFRIENDLY DOMINANT LIGHT FUNDAMENTAL PARTICLE ON EARTH.

LIGHT ENERGY –CONTENT OF ONE GREEN COLOR LIGHT PARTICLE (Y. g. –F.P.) IS EQUAL TO ONE UNIT LIGHT – ENERGY- AMOUNT,

THE DIFFERENT LIGHT FUNDAMENTAL PARTICLES HAVE DIFFERENT LIGHT – ENERGY CONTENTS, DIFFERENT PARTICLE WEIGHTS, DIFFERENT GRAVITON AS WELL AS DIFFERENT PHYSICAL CHEMICAL BIOLOGICAL PROPERTIES, THE LIGHT ENERGIES FROM DIFFERENT SOURCES CAN BE QUANTIZED AND MEASURED IN COMPARE TO ONE UNIT LIGHT-ENERGY THAT DEFINED IN ABOVE, THE YELLOW COLOR LIGHT FUNDAMENTAL PARTICLE (Y. y.- F.P.) IN EARTH IS ANOTHER EXAMPLE FROM SENSIBLE BIOFRIENDLY LIGHT PARTLE CLASSES, THERE ARE LARGE NUMBERS OF OTHER NON DISCOVERED LIGHT FUNDAMENTAL PARTICLE CLASSES IN EARTH WHICH HAVE BEEN USED IN CONSTRUCTIONS OF DIFFERENT NANO-UNITS.

UNIT OF ELECTRIC – ENERGY

ELECTRIC ULTRAVIOLET FUNDAMENTAL PARTICLE (E. UV- FP)

THE EXISTING ELECTRIC ENERGY AMOUNT IN ONE ULTRAVIOLET ELECTRIC PARTICLE IS EQUALTO ONE UNIT OF ELECTRIC ENERGY.

AN ULTRAVIOLET FUNDAMENTAL PARTICLE (E. UV - FP) IS THE WEAKEST ELECTRIC FUNDAMENTAL PARTICLE BETWEEN EARTH'S ELECTRIC FUNDAMENTAL PARTICLES, AN ULTRAVIOLET PARTICLE POSSESSES THE LEAST PARTICLE WEIGHT, PARTICLE GRAVITON AND ELECTRIC ENERGY CONTENT IN COMPARISON TO OTHER EARTHS

THE ELECTRIC ENERGY UNIT: THE ELECTRIC ENERGY AMOUNT OF ONE ULTRAVIOLET FUNDAMENTAL PARTICLE IS EQUAL TO THE ONE UNIT ELECTRIC ENERGY QUANTITY.

THE ULTRAVIOLET PARTICLE IS LOCATED, AFTER THE FINISHING LINES OF LIGHT FUNDAMENTAL PARTICLE CLASS PARTICLE GROUPS, AND THE ULTRAVIOLET PARTICLE'S LOCATION IS, JUST BEFORE THE STRATING POINTS OF HARSH BIOHOSTILE HIGH ENERGY INDUSTRIAL ELECTRIC FUNDAMENTAL PARTICLE CLASSES.

QUANTUM ELECTRIC-ENERGY PROVIDERS FOR INTER ELECTRON-NUCLEON NANO- SITES OF LIVING THINGS

VIOLET COLOR ELECTRIC PARTICLES (EV – FP), BLUE COLOR ELECTRIC PARTICLES (EB– FP)

BLUE COLOR LIGHT PARTICLES (YB-FP) AND VIOLET COLOR LIGHT PARTICLE (YV-FP)

VIOLET COLOR PARTICLES (Y. v. – F.P.) AND BLUE COLOR FUNDAMENTAL PARTICLES (Y. b. – F.P.) POSSESSES MIXED LIGHT AND ELECTRIC ENERGIES, BOTH PARTICLES ARE LIGHT ENERGY AND ELECTRIC ENERGY PROVIDER INTO INTERNAL ELECTRON - NUCLEON NANO-SITES TO BE USED FOR ACHIEVING QUATUM BIOLOGICAL PHYSICAL

CHEMICAL FUNCTIONS, IN EARTH THERE ARE LARGE OTHER UNKNOWN NON SENSIBLE MULTI ENERGY PROVIDER FUNDAMENTAL PARTICLES, WHICH CONSTRUCT DIFFERENT INTERNAL ELECTRON – NUCLEON PARTICLE COMPOUNDS CONSTRUCTIONS OF ATOMS AND PRESENTLY REMAIN UNKNOWN.

BLUE COLOR PARTICLES, VIOLET COLOR PARTICLES ARE BIOFRIENDLY LIGHT - ELECTRIC PARTICLES, IN LIVING THINGS THESE PARTICLES PROVIDE ELECTRIC ENERGIES AND LIGHT ENERGIES INTO INTERNAL ELECTRON NUCLEON QUATUM LOCATIONS, WITH COOPERATION OF OTHER BIOFRIENDLY FUNDAMENTAL PARTICLES OPERATE ENTIRE LIVING THINGS BIOLOGICAL, CHEMICAL, PHYSICAL NANO FUNCTIONS IN DIFFERENT QUANTUM LOCATIONS.

THE UNIVERSE UNDER THE GRAVITON LAWS AND ORDERS

THE GRAVITON FORCE IS THE PROPERTY OF THE MASSES

GRAVITON ALSO HAVE POSSIBILITIES FOR BEING A DIFFERENT CLASSES OF THE FUNDAMENTAL PARTICLES, WHICH POSSESS REMARKABLE ATTRACTIVE AND REPULSIVE FORCE PROPERTIES, WITH TWO MAJOR CLASS GRAVITON PARTICLES, ONE FORCES IS OPPOSITES OF THE OTHERS.

THE CHEMICAL COMBINATIONS AND THE AUTONOMOUS SEQUENTIAL CHEMICAL CHAINS INTERACTIONS BETWEEN THE DIFFERENT MASSES:

UNDER ATTRACTIVE GRAVITON FORCES TWO COMPATIBLE MASSES ATTRACT EACH OTHER AND COMBINE.

UNDER REPULSIVE GRAVITON FORCES ORDERS TWO NON –COMPATIBLE MASSES REJECT AND REFUSE COMBINATIONS.

THE ABOVE LAWS AND ORDERS START FROM PARTICLE SIZE MASSES, UP TO UNIVERSAL MASS SIZES WITH NO DIFFERENCE IN RULES.

UNDER ATTRACTIVE GRAVITON FORCE ORDERS REGENERATIVE AUTONOMOUS SEQUENTIAL CHEMICAL INTERACTIONS CHAINS AND CYCLES TAKE PLACE AND MASSES COMBINE TO EACH OTHER, BUT UNDER REPULSIVE GRAVITON ORDERS THE DEGENERATIVE AUTONOMOUS SEQUENTIAL CHEMICAL INTERACTIONS CYCLES CAUSE BREAK DOWN OF LARGE CONSTRUCTIONS, AND MASSES REFUSE COMBINATIONS,

THE ABOVE RULE APPLIES TO ALL OTHER AUTONOMOUS SEQUENTIAL CHEMICAL INTERACTION CYCLE BETWEEN ATOM-BASE OR NANO-UNIT-BASE OR CELL AND MICRO-UNIT BASE, AS WELL TO MACRO-UNIT AUTONOMOUS SEQUENTIAL CHEMICAL INTERACTION CHAINS WITH NO EXCEPTIONS,

THE GRAVITON CONTROL OF MASSES ALSO APPLY TO MICRO-UNIT STRUCTURES AS WELL. UNDER GRAVITON, ATTRACTIVE OR REPULSIVE FORCES ORDERS EITHER TWO MICRO-UNIT CHROMOSOMAL STRUCTURES OR TWO CELLS, THEY EITHER COMBINE OR REFUSE COMBINATIONS.

THE COMBINATION OR REPULSION OF LARGE MASSES SUCH AS PLANETS, GALAXIES, AS WELL AS UNIVERSES ARE ALSO ACHIEVE UNDER GRAVITON FORCES DIRECTLY OR EITHER COMBINED OR NOT, THAT IS THE ORDERS AND LAWS OF GRAVITON ACROSS THE UNIVERSE OVER THE MASSES.

UNIT OF GRAVITON

=====================

THE EXISTING GRAVITON ENERGY FORCE QUANTITY IN ONE EARTH'S AIRBORN LIGHT FUNDAMENTAL PARTICLE'S GRAVITY ENERGY, IS EQUALS TO ONE UNIT GRAVITON.

THE EXISTING ATTRACTIVE OR REPULSIVE GRAVITY ENERGY FORCE IN ONE GRAVITON PARTICLE IS EQUAL TO ONE UNIT GRAVITON FORCE.

RULES, LAWS AND ORDERS OF THE GRAVITON ACROSS THE UNIVERSE

NUCLEUS GENESIS & PERIPHERON GENESIS PHENOMENONS UNDER ORDERS OF GRAVITONS.

THE GRAVITON IS ENERGY FORCE IN MASS CONTENT, THE MASS WHO POSSESS MORE GRAVTON, IT IS POWERFUL AND DOMINANT, THROUGH GRAVITON ENERGY FORCES THE MASSES EITHER ATTRACT EACH OTHER AND COMBINE, OR UNDER REPULSIVE GRAVITON ENERGY FORCES MASSES REFUSE COMBINATIONS AND STAY APPART.

THERE IS DIRECT RELATIONSHIPS BETWEEN MASS QUANTITY WITH GRAVITON QUANTITY.

THE GRAVITON – UNIT: THE AMOUNT OF GRAVITON ENERGY IN ONE WHITE LIGHT PARTICLE IS EQUAL TO ONE UNIT GRAVITON FORCE.

THE AMOUT OF GRAVITON ENERGY FORCES (EITHER THE ATTRACTION GRAVITY FORCE ENERGY AMOUNT, OR THE REPULSION GRAVITY

ENERGY QUANTITY) QUANTITY IN ONE GRAVITON FUNDAMENTAL PARTICLE IS EQUAL TO ONE GRAVITON ENERGY-UNIT.

THE MASS INTERACTIONS WITH EACH OTHER UNDER GRAVITON ENERGIES ORDERS

1 -THE MOLECULES COMBINE TO EACH OTHER UNDER ATTRACTIVE GRAVITON FORCES AND PRODUCE MOLECULES.

2 - THE REPULSIVE GRAVITON FORCES PREVENT FROM COMBINATIONS, THROUGH MASSES REPULSIVE GRAVITON FORCES, CAUSE SEPARATION OF MOLECULES FROM EACH OTHER.

3 - GRAVITON ORDERS APPLY TO ALL MASS -UNITS SIZES, STARTING FROM FUNDAMENTAL PARTICLE MASS –UNITS, ATOM-BASE MASS UNITS, MICRO-UNITS, MACRO-UNITS, UP TO THE UNIVERSAL SIZE MASSES.

4 - UNDER GRAVITON FORCES THE NANO-UNITS, MICRO –UNITS AND MACRO-UNIT MASSES EITHER COMBINE UNDER ATTRACTIVE GRAVITON FORCES, OR STAY APART BY REPULSIVE GRAVITON ENERGY ORDERS.

5 - THE LESS WEIGHT MASS UNITS POSSESSES LESS GRAVITON ENERGY FORCES, THE LESS WEIGHT MASS- UNITS HAVE LESS ENERGY CONTENTS OF ANOTHER KIND ENERGIES.

6 - THE HIGH WEIGHT MASS UNITS POSSESSES HIGH GRAVITON ENERGY FORCES AND HIGH ENERGY CONTENTS OF OTHER KIND ENERGIES RELATIVELY.

7 - THE HIGH WEIGHT MASS UNITS DOMINATE AND CONTROL THE LOW WEIGHT MASS –UNITS.

8 – THE DOMINANT MASS UNITS WITH HIGH GRAVTON ENERGY FORCES, MOSTLY TAKE CENTRAL LOCATION AND CONSTRUCT A NUCLEUS WITH HIGH ENERGY GRAVITON FORCES, THE DOMINANT MASSES THROUGH HIGH ATTRACTIVE GRAVITON FORCES FROM NUCLEUS ATTRACT RECESSIVE MASS UNITS IN THEIR PERIPHERY, AND

COMBINE ALWAYS THE DOMINANT MASS UNITS STAY IN CENTER, AND RECESSIVE MASS - UNIT ALWAYS STAY IN PERIPHERONS.

9 – IN COMBINATION OF MASSES THE RECESSIVE LESS WEIGHT, LESS GRAVITON, LESS ENERGY NANO-UNIT STAY IN PERIPHERY AND CONSTRUCT PERIPHERON (THE ELECTRONS), BUT THE DOMINANT HIGH WEIGHT, HIGH ENERGY, HIGH GRAVITON NUCLEUS NANO-UNITS (THE NUCLEONS), ALWAYS BUILD A NUCLEUS, WHICH THE NUCLEUS ALWAYS TAKE CENTRAL POSITIONS AND FROM IT'S NUCLEAR LOCATION THROUGH HIGH GRAVITON ENERGY ATTRACT THE RECESSIVE LESS GRAVTON ENERGY ELECTRONS MASS UNITS.

10 – THE ABOVE RULES OF GRAVITON IS TRUE IN REGARD TO FUNDAMENTAL PARTICLES ALSO, ALL OVER THE UNIVERSE IN ALL PLANETS ACROSS THE UNIVERSE.

11 - IN COMBINATION OF LESS MASS UNIT HYDROGEN, WITH HIGH MASS UNIT OXYGEN, THE DOMINANT GRAVITON FORCE OXYGEN TAKES CENTRAL NUCLEUS POSITION AND FROM THERE BY ATTRACTIVE GRAVITON ENERGY FORCES ATTRACT LESS WEIGHT, LESS ENERGY GRAVITON FORCE TWO HYDROGEN ATOMS, WHICH THE LESS WEIGHT HYDROGENS STAY IN PERIPHERY ORBITS, UNDER OXYGENS DOMINANT GRAVITON ORDERS AND CONSTRUCT PERIPHERON, AND OXYGEN STAY IN CENTER AND FROM NUCLEI POSITION CONTINUOUSLY CONTROL HYDROGEN.

12 – THE ABOVE LAWS AND RULES ORDERS OF GRAVITON IS TRUE IN REGARD TO MICRO – UNITS MASSES AS WELL, WHEN A DOMINANT HIGH WEIGHT, HIGH GRAVITON FORCE CELLS - NUCLEUS COMBINE WITH RECESSIVE LESS WEIGHT, LESS ENERGY GRAVITON FORCE CYTOPLASMA, ALWAYS THE CELL NUCLEUS TAKE CENTRAL POSITION UNDER DOMINANT HIGH FORCE GRAVITON ORDERS AND THE CYTOPLSMA WITH LESS GRAVITON FORCE ENERGY OBEY AND STAY AT PERIPHERY.

13 - ABOVE RULES ARE TRUE IN MACRO-UNITS MASSES AND UNIVERSAL SIZE MASS SYSTEMS AS WELL, THE GRAVITON RULES FOR ALL SIZES DIFFERENT MASSES STAY THE SAME.

IN REGARD TO UNIVERTSAL MASS SYSTEMS, THE LARGE CENTRAL UNIVERSE PLANETARY MASS SYSTEMS (THE UNIVERSAL – NUCLEI), FROM NUCLEUS POSITION THROUGH IT'S POWERFUL GRAVITON ATTRACTIVE FORCES CONTROL ALL OTHER PERIPHERAL SMALLR MULTI UNIVERSE MASS SYSTEMS OF UNIVERSAL PERIPHERON (THIS PERIPHERAL MASS SYSTEMS CALLED AS: UNIVERSAL – PERIPHERON), WHICH EACH GIVEN MASS – UNIT POSSESSES FIXED POSITION AND STAY RELATIVELY IN CERTAIN FIXED LOCATION IN REGARD TO OTHER MASSES UNDER THE FORCES OF GRAVITON - ORDERS.

THE MASS UNIT OR THE UNIT OF THE WEIGHT

ONE GREEN COLOR LIGHT FUNDAMENTAL PARTICLE'S WEIGHT AT EARTH IS EQUAL TO ONE UNIT MASS

(ONE MASS UNIT IS EQUALS TO WEIGHT OF ONE Y. g. – F.P. AT EARTH LOCATION)

THE OTHER FUNDAMENTAL PARTICLES WEIGHTS QUANTIZED IN THE COMPARISON TO THE WEIGHT OF ONE GREEN LIGHT PARTICLE.

THE DIFFERENT FUNDAMENTAL PARTICLES (EVEN FROM THE SAME CLASS) IN OTHER PLANETS POSSESSES DIFFERENT WEIGHTS.

THE WEIGHTS, ENERGIES, GRAVITON FORCES OF ALL DIFFERENT CLASSES OF FUNDAMENTAL PARTICLES IN HARSH PLANETS, SUCH AS BLACK HOLES, ARE THOUSANDS FOLDS HIGHER THAN THE EARTH. AND THE CONSTRUCTED NANO-UNITS FROM CHEMICAL COMBINATIONS OF THESE FUNDAMENTAL PARTICLES PRODUCE ATOMS SUCH AS BLACK MATTER, DIFFERENT ELECTRONS, NUCLEONS CONSTRUCTIONS.

THE ENERGY UNITS

1 - ONE UNIT OF THERMAL ENERGY IS EQUAL TO THE HEAT ENERGY QUANTITY OF ONE INFRA- RED –PARTICLE ENERGY AMOUNT IN EARTH.

2 - ONE UNIT OF ELECTRIC –ENERGY IS EQUAL TO THE ELECTRIC ENERGY CONTENT OF ONE ULTRA VIOLET PARTICLE ENERGY AMOUNT ON EARTH.

3 - ONE UNIT OF LIGHT ENERGY IS EQUAL TO THE LIGHT ENERGY AMOUNT OF ONE GREEN COLOR PARTICLE ENERGY QUANTITY AT PLANET EARTH.

4 - CONSIDERING THE EARTH'S OTHER CLASSES OF FUNDAMENTAL PARTICLES, WHICH THEY HAVE ANOTHER KIND ENERGY CONTENTS, ALL ARE CONSIDERED TO BE ONE UNIT ENERGY QUANTITY. FOR EXAMPLE, THE GAMMA ENERGY UNIT, IS EQUAL TO ONE GAMMA PARTICLE ENERGY QUANTITY IN ONE GAMMA PARTICLE AT PLANET EARTH REGION. OR ONE X –ENERGY UNIT, IS EQUAL TO ONE X- PARTICLE'S ENERGY QUANTITY CONTENT, IN ONE X-FUNDAMENTAL PARTICLE CONTENT IN PLANET EARTH REGION, ETC.

THERE ARE LARGE NUMBERS OF OTHER PLANETS AND GALAXIES WHICH THEIR FUNDAMENTAL PARTICLES ENERGY –CONTENTS ALL ARE BIO-LETHAL AND POWERFUL ELECTRIC, THERMAL, LIGHT, SONIC, GAMMA, X. FUNDAMENTAL PARTICLES WHICH THEIR ENERGY- CONTENTS AND ENERGY INTENSITIES ARE THOUSANDS TO MILLIONS FOLDS HIGHER THAN THE EARTH'S PARTICLES.

FUNDAMENTAL-PARTICLE COLONY

FUNDAMENTAL PARTICLES TEND TO COLONIZE AND TRAVEL IN WAVE SHAPES. THE TOTAL FUNDAMENTAL PARTICLE-POPULATION NUMBERS IN EACH GIVEN ONE WAVE-LENGTH IS ONE FUNDAMENTAL PARTICLE-COLONY. FOR EACH GIVEN FUNDAMENTAL PARTICLE THE WAVE LENGTH, FREQUENCY, FUNDAMENTAL PARTICLE COLONY AND TOTAL FUNDAMENTAL PARTICLE POPULATION NUMBER PER COLONY ARE WELL DEFINED, CONSTANT RELATIVELY FIXED, CERTAIN FOR ANY GIVEN FUNDAMENTAL PARTICLE.

THE TOTAL PARTICLE -POPULATION NUMBER AT EACH GIVEN WAVE –LENGTH IS EQUAL TO ONE FUNDAMENTAL-PARTICLE COLONY, PARTICLE COLONY FOR ANY GIVEN PARTICLE ALWAYS IS FUNDAMENTAL PARTICLES SPECIFIC AND DIFFERENT FUNDAMENTAL PARTICLE CLASSES POSSESSES DIFFERENT FUNDAMENTAL PARTICLES COLONIES.

HOMOGENEOUS PARTICLE COLONIES THE TOTAL PARTICLE- POPULATION AT ONE WAVE –LENGTH AND ARE ALL COMPOSED FROM ONE KIND OF FUNDAMENTAL PARTICLE CLASS HOMOGENOUSLY,

AT HETEROGENEOUS FUNDAMENTAL PARTICLE COLONIES, THERE ARE DIFFERENT KINDS OF FUNDAMENTAL PARTICLES IN ONE WAVE LENGTH COLONY HETEROGENEOUSLY CONSTRUCTING PARTICLE COLONIES. IN HETEROGENEOUS PARTICLE COLONIES, THERE ARE MANY KIND DIFFERENT FUNDAMENTAL PARTICLES FROM DIFFERENT CLASSES IN ONE WAVE LENGTH COLONY.

THE PLANET EARTH'S FUNDAMENTAL PARTICLES (F.P.)
AND

THE FUNDAMENTAL PARTICLE CLOUDS (FP – Cl)

THE FUNDAMENTAL PARTICLE

INFORMATION AND IMAGE PARTICLE – CLOUDS

(F.P. – I. I. –P. cl.)

THE MOST OF THE FUNDAMENTAL PARTICLES AT EARTH'S ATMOSPHERE, ARE LESS ENERGY, LOW GRAVITON FORCE, NON DISCOVERED, BIO FRIENDLY, LOW WEIGHT FUNDAMENTAL PARTICLES (FP), AND THE FUNDAMENTAL PARTICLE CLOUDS (FP – Cl.), THESE FUNDAMENTAL PARTICLES AND OTHER KNOWN AND UNKNOWN OTHER CLASSES OF THE FUNDAMENTAL PARTICLES ALL TOGETHER CONSTRUCT THE EARTH'S ELECTRON, NUCLEON, ATOM AND NANO – UNIT PARTICLE COMPOUND'S MOLECULAR CONSTRUCTIONS. AND FROM THE KNOWN BIO FRIENDLY FUNDAMENTAL PARTICLE CLASSES OF THE ATOMSPHERE, ONLY A FEW SENSABLE CLASSES OF THE FUNDAMENTAL PARTICLES HAVE BEEN DISCOVERED TODAY.

THE POTENT HEAVY WEIGHT, HIGH PARTICLE MASS, HIGH GRAVITON FORCE, HIGH ENERGY CONTENT, THE BIO LETHAL FUNDAMENTAL PARTICLE CLASSES ARE ABSENT MOSTLY FROM ATMOSPHERE OF PLANET EARTH. THE DIFFERENT CLASS PARTICLES, WITH DIFFERENT KIND ENERGY TYPES, SPECIFICALLY WITH HIGH ENERGY CONTENT PARTICLES SUCH AS; GAMMA PARTICLE CLASSES FUNDAMENTAL PARTICLES, WITH HIGH GAMMA ENERGY, OR BIO CIDAL X – FUNDAMENTAL PARTICLES WITH X – ENERGY CONTENT PARTICLES, OR THE BIO HOSTILE POTENT HIGH SONIC ENERGY SOUND FUNDAMENTAL PARTICLES THAT THEIR PARTICLES ARE CAPABLE TO TRAVEL SEVERAL THOUSAND LIGHT YEARS, OR LETHAL POWER LIGHT FUNDAMENTAL PARTICLES, ELECTRIC PARTICLES, OR THE KIND OF THERMAL FUNDAMENTAL PARTICLES THAT THEIR HEAT ENERGY CONTENTS EXCEED SEVERAL MILLION DEGREES CENTIGRADE PARTICLES, ETC. MOSTLY NOT EXISTED IN EARTH'S ATMOSPHERE AND SURFACE AT LEAST.

EARTH'S KNOWN BIO FRIENDLY FUNDAMENTAL PARTICLES AND PARTICLE CLOUDS

BIO FRIENDLY LIGHT FUNDAMENTAL PARTICLES, LIGHT PARTICLE'S INFORMATION AND IMAGE CLOUDS.

(Y-FP – I.I. – P. Cl.)

THERE ARE DIFFERENT SUB - CLASSES FROM LIGHT- FUNDAMENTAL PARTICLES, LIGHT FUNDAMENTAL PARTICLE CLOUDS (Y. - F.P. - Cl.), AND DIFFERENT CLASSES OF LIGHT FUNDAMENTAL PARTICLES INFORMATION – IMAGE

PARTICLE CLOUDS (Y. –FP. – I.I. – P. Cl.) IN EARTH, WHICH ALL OF THESE LIGHT FUNDAMENTAL PARTICLES IN PLANET EARTH ARE LOW FUNDAMENTAL PARTICLE WEIGHT, LESS GRAVITON FORCE CONTENT, LOW LIGHT – ENERGY QUANTITY, WEAK ENERGY AND NON BIO - LETHAL, LIGHT FUNDAMENTAL PARTICLES, LIGHT FUNDAMENTAL PARTICLE CLOUDS, AND LIGHT FUNDAMENTAL PARTICLE INFORMATION AND IMAGES PARTICLE CLOUDS (Y – FP –

I.I. – P. Cl.), EITHER WITH SIMPLE OR COMPLEX MOLECULAR STRUCTURES, MAJORITY OF Y – FP OF THE EARTH PRESENTLY HAVE NOT BEEN DISCOVERED YET, ONLY WE KNOW A FEW SENSIBLE LIGHT FUNDAMENTAL PARTICLE CLASSES FROM LIGHT FUNDAMENTAL PARTICLES (Y. – F.P.), WHICH SOME OF THOSE CLASSES ARE SUCH AS:

THE RED COLOR LIGHT FUNDAMENTAL PARTICLE (Yr. – F.P.) SUBCLASSES, THE RED FUNDAMENTAL PARTICLE CLOUDS (Yr. – F.P. – Cl.), AND RED COLOR LIGHT FUNDAMENTAL PARTICLE INFORMATION IMAGE PARTICLE CLOUDS (Yr. – F.P. – I.I. – P. Cl.) HAVE PLAYED ESSENCIAL ROLES IN PRODUCTION OF LIVING THINGS AND LIVE NANO - UNITS, THE RED COLOR PARTICLES POSSESS ALSO THERMAL ENERGY, THEY PROVIDE QUANTUM HEAT ENERGY FOR NANO-LOCATIONS, THE NANO ENERGIES OF DIFFERENT KNIDS ARE NEEDED FOR PERFORMING DIFFERENT PHYSICAL, CHEMICAL, BIOLOGICAL NANO- TASKS INSIDE ELECTRONS, NUCLEONS AND OTHER NANO-UNITS LOCATIONS.

THE YELLOW COLOR LIGHT FUNDAMENTAL PARTICLE (Y. y. – F.P.) SUBCLASSES, THE YELLOW COLOR LIGHT FUNDAMENTAL PARTICLE CLOUDS (Y. y. – F.P. – Cl.), AND YELLOW COLOR LIGHT FUNDAMENTAL PARTICLE INFORMATION AND IMAGE PARTICLE CLOUDS (Y. y. – F.P. – I.I. – P. Cl.) ALL ARE DIFFERENT CLASSES FROM BIO FRIENDLY LIGHT PARTICLES,

THE GREEN COLOR LIGHT FUNDAMENTAL PARTICLES (Y. g. – F.P.) SUBCLASSES, THE GREEN LIGHT FUNDAMENTAL PARTICLE CLOUDS (Y. g. – FP – Cl), AND GREEN LIGHT FUNDAMENTAL PARTICLES INFORMATION IMAGE PARTICLE CLOUDS (Y. g. – F.P. – I.I. – P. Cl.), ALL ARE LIGHT ENERGY PROVIDERS, POSSESS QUANTUM LIGHT - ENERGY CONTENT, THE MOLECULAR GREEN PARTICLE COMPOUNDS ARE VITAL FOR PLANTS AND LIVING THINGS, THE GREEN LIGHT FUNDAMENTAL PARTICLES CONSTRUCT THE INTERNAL ELECTRON – NUCLEON PARTICLE COMPOUND CONSTRUCTIONS OF THE PLANTS, AND ARE ESSENTIAL IN PLANT'S ELECTRON – GENESIS AND NUCLEON -GENESIS.

THE BLUE COLOR LIGHT FUNDAMENTAL PARTICLE (Y. b. – F.P.) CLASSES, AND THE VIOLET COLOR LIGHT FUNDAMENTAL PARTICLES (Y. v. – F.P.) BOTH POSSESS QUANTUM ELECTRICAL ENERGY CONTENTS, AND ARE BIO FRIENDLY ELECTRIC ENERGY PROVIDERS FOR NAO-SITES OF THE INTERNAL ELECTRON – NUCLEON LOCATIONS, THE BIO FRIENDLY DIFFERENT KIND ENERGY CONTENT FUNDAMENTAL PARTICLES OPERATE LIVING THINGS INTERNAL NANO LOCATIONS PHYSICAL, CHEMICAL, BIOLOGICAL NANO – TASKS, THROUGH PROVIDING DIFFERENT KIND NEEDED NANO – ENERGIES FOR NANO- LOCATIONS,

THE BLUE COLOR, THE VIOLET COLOR FUNDAMENTAL PARTICLES WITH OTHER NON DISCOVERED ELECTRIC FUNDAMENTAL PARTICLES CLASSES, ALL ARE ESSENTIAL FOR CONTINUATION OF LIFE AND EXISTENCES OF THE LIVING THINGS, WITHOUT BIO FRIENDLY ELECTRIC FUNDAMENTAL PARTICLES THE CONTINUATIONS OF THE LIVING THINGS LIVES ARE NOT POSSIBLE, AND THEIR LIVES EXISTENCE SEASE IN MATTERS OF THE SECONDS SPECIALLY IN ANIMAL SPECIES.

IN EARTH, THERE ARE NUMEROUS OTHER CLASSES OF BRILLIANT EXCEPTIONALLY BRIGHT OTHER COLORS OF THE SENSIBLE LIGHT FUNDAMENTAL PARTICLE SUBCLASSES WHICH ARE NOT SEEN AND DISCOVERED YET AT PRESENT TIMES, THESE LIGHT PARTICLE ARE USED IN CONSTRUCTIONS OF THE INTERNAL ELECTRON – NUCLEON LIGHT FUNDAMENTAL PARTICLE COMPOUNDS OF THE DIFFERENT ATOMS, SPECIALLY IN BUILDING OF LIGHT PARTICLE COMPOUNDS, THESE LIGHT PARTICLES ARE NOT AIRBORN AND HAVE NOT BEEN DISCOVERED AS FREE LIGHT PARTICLE.

MANY OF THESE TRAPPED NON –FREE SENSIBLE LIGHT PARTICLES, ARE STORED INSIDE LIGHT PARTICLE CONSTRUCTED ATOMS, AND BUILDING THE LIGHT PARTICLE COMPOUND CONSTRUCTIONS OF LIGHT – PARTICLE – ATOMS ELECTRONS AND NUCLEONS, THESE LIGHT PARTICLES CAN BE EASILY RETRIEVED TO OUT OF ELECTRONS NUCLEONS UNDER DEGENERATIVE A. S. I. F.P. Mol. C.I.C.

THE SONIC FUNDAMENTAL PARTICLES (S – FP)

IN EARTH, MOST SOUND – FUNDAMENTAL PARTICLE SUBCLASSES ARE LOW PARTICLE WEIGHT, LOW SONIC – ENERGY, LOW GRAVITON FORCE, BIO FRIENDLY PARTICLES, SOUND PARTICLE CLOUDS (S – FP- Cl), AND SONIC FUNDAMENTAL PARTICLE INFORMATION IMAGE PARTICLE CLOUDS (S – FP – I.I. – P. Cl). AND MANY SOUND PARTICLE CLASSES YET NOT DISCOVERED AND ARE UNKNOWN.

BIO FRIENDLY THERMAL FUNDAMENTAL PARTICLES (T – FP)

THE THERMAL PARTICLES, THERMAL FUNDAMENTAL PARTICLE CLOUDS (T. – F.P. - Cl.), AND THERMAL FUNDAMENTAL PARTICLE INFORMATION AND IMAGE PARTICLE CLOUDS (T – FP – I.I. – P. Cl.), ARE CONSTRUCTION ELEMENTS OF ELECTRON, NUCLEON PARTICLE COMPOUNDS CONSTRUCTIONS, AND ARE THERMAL ENERGY PROVIDERS FOR INTERNAL ATOM NANO- FUNCTIONS, PLAY ESSENTIAL ROLES IN PERFORMING PHYSICAL, CHEMICAL, BIOLOGICAL TASKS OF LIVING THINGS, THE INFRA – RED FUNDAMENTAL PARTICLES (Yir – FP = T ir - FP) CLASSES ARE ONE EXAMPLE FROM BIO FRIENDLY THERMAL PARTICLES.

BIO FRIENDLY ELECTRIC FUNDAMENTAL PARTICLES (E. - F.P.)

THE BIO FRIENDLY ELECTRIC FUNDAMENTAL PARTICLES (E – FP), ELECTRIC FUNDAMENTAL PARTICLE CLOUDS (E – FP – Cl), THE ELECTRIC FUNDAMENTAL PARTICLES INFORMATION IMAGE PARTICLE CLOUDS (E – FP – I.I. –P. Cl.) ARE VITAL FOR PRODUCTIONS OF LIVING THINGS, AND CONTINUATIONS OF LIFE IN EARTH, THESE PARTICLE CLASSES MOSTLY ARE NOT DISCOVERED YET, IN EARTH WITHOUT ELECTRIC FUNDAMENTAL PARTICLE LIFE CAN NOT EXIST, ULTRA – VIOLET FUNDAMENTAL PARTICLES ARE FROM BIO HOSTILE ELECTRIC FUNDAMENTAL PARTICLE CLASSES OF THE PLANET EARTH.

THERE ARE LARGE NUMBERS OF OTHER NON - DISCOVERED, NON SENSIBLE, BIOFRIENDLY FUNDAMENTAL PARTICLES CLASSES IN EARTH'S ATMOSPHERE, AND THESE PARTICLES CONSTRUCT MOST OF THE PLANET EARTHS ELECTRON- NUCLEON PARTICLE COMPOUND CONSTRUCTIONS. THE FREE AIRBORN BIO-LETHAL FUNDAMENTAL PARTICLE CLOUDS SUCH AS, X- PARTICLES, GAMMA PARTICLES ARE RARE IN EARTH'S ATMOSPHERE. THE OTHER PLANETS PARTICLE CLOUDS MOSTLY ARE BIOHOSTILE, AND ARE NOT LIVABLE.

THE ATOM - CLOUDS, ATOM INFORMATION-IMAGE CLOUD, NANO – UNIT INFORMATION IMAGE - CLOUD (N. U. - I. I. - Cl.), MICRO –UNITS, CELLS, AND MICRO-UNIT CLOUDS, ETC., ARE ANOTHER BIO-FRIENDLY OR BIOHOSTILE STRUCTURES OF PLANET EARTHS ATMOSPHERE CONTENTS.

BIO HOSTILE FUNDAMENTAL PARTICLES

THE BIO HOSTILE FUNDAMENTAL PARTICLES SUCH AS: X – FUNDAMENTAL PARTICLES, GAMMA FUNDAMENTAL PARTICLES CLASS, BIO – CIDAL ELECTRIC FUNDAMENTAL PARTICLES CLASSES OR LETHAL SONIC FUNDAMENTAL PARTICLE CLASSES, BIO – LETHAL THERMAL FUNDAMENTAL PARTICLES CLASSES, ETC. NOT EXPLAINED IN THIS BOOK.

THE MIXED NANO CLOUDS, MICRO CLOUDS OF EARTH'S ATMOSPHERE

THE EARTH'S ATMOSPHERE COMPOSED FROM FLOATING NANO UNITS SUCH AS: ELECTRONS, PROTONS, NEUTRONS, ATOMS, ATOM BASE MOLECULAR STRUCTURES, ATOM CLOUDS, ELECTRON CLOUDS, NUCLEON CLOUDS, AND ANOTHER KINDS OF NANO – UNIT CLOUDS, NANO- UNIT INFORMATION IMAGE CLOUDS (NU – I.I. – Cl.), FUNDAMENTAL PARTICLES, FUNDAMENTAL PARTICLE CLOUDS (FP – Cl.), FUNDAMENTAL PARTICLE INFORMATION IMAGE PARTICLE CLOUDS (F.P. – I.I. – P. Cl.), ATOM BASED INFORMATION IMAGE CLOUDS (A. – I.I. –

Cl.), FLOATING MICRO UNITS, AND MICRO - UNIT CLOUDS, MICRO UNIT INFORMATION IMAGE CLOUDS, AND FLOATING DIFFERENT SIZES MACRO – UNIT MASSES IN PLANET EARTH'S ATMOSPHERE.

EXAMPLES FROM NANO SIZE MASS – UNITS (NANO – UNITS) ARE THE MASS SUCH AS: FUNDAMENTAL PARTICLES, ELECTRONS, NEUTRONS, ATOMS.

MICROSCOPIC SIZE MASS UNITS (MICRO – UNITS) EXAMPLES ARE SUCH MASS AS: CELLS, CHROMOSOMES, MITOCHONDRIA, ETC., MACRO SIZES MASS UNITS ARE THE SENSIBLE MASSES OF ANY SIZE, UP TO GLOBAL.

FUNDAMENTAL PARTICLE WEAPONS OF MASS DESTRUCTION, AND CRIMES AGAINST HUMANITY

SUPER-SONIC PARTICLE WEAPON'S INJURIES

MICROWAVE WEAPONS OF MASS DESTRUCTION AND MICROWAVE WEAPON WOUNDS,

CRIMES AGAINST HUMANITY

SHOOTING MICROWAVE SUPER-SONIC LIGHT CODED LASER -PARTICLES WEAPONS OF MASS DESTRUCTION, CAN BE FIRED AIRBORN FROM ANY DISTANCE, FROM ROUGES ARMY BASE, OR FROM POWERFUL TELECOMMUNICATION TOWERS, THROUGH COMPUTERIZED WELL CALIBRATED AND DOSE ADJUSTED VARIABLE LETHAL AND NON-LETHAL MICROWAVE DOSES DELIVERED COMPUTERIZED INTO ANY SELECTED HOUSES IN ANY CITY. THE ROUGES AGENTS CAN SHOOT OR TORTURE BY MICROWAVE ANYONE INSIDE THEIR OWN HOMES WITHOUT LEAVING ANY TRACE, THE POLICE IS PART OF THE GANG. USE OF THESE DEVICES DO NOT LEAVE TRACE FROM KILLING IN BEHIND.

MICROWAVE SUPERSONIC WEAPONS INJURIES ARE INTERNAL ELECTRON-NUCLEON PARTICLE COMPOUND CONSTRUCTIONS, THE PARA- CLINICAL STUDIES SUCH AS CHEMICAL STUDIES, MRI, CAT SCANS, PET SCAN OR SIMILAR STUDIES MAY NOT HELP? WE MAY HAVE SOME IN THE FUTURE, THESE ABUSES ARE COMMON IN LESS ADVANCED REGIONS AT MOST.

THE INTERESTING FACT IS, THAT INTERNAL ATOM INJURIES BY MICROWAVE PRODUCE PATIENT'S COMPLAINTS, SIMILAR TO COMPLAINTS OF INJURIES CAUSED BY OTHER REASONS AND DISEASES, AND THE HISTORY TRACE IS SOME THING SIMILAR AS ABOVE, BUT THERE IS NO VISIBLE PHYSICAL EVIDENCES.

HEAD INJURIES

DESTRUCTION OF INTERNAL ELECTRON'S NUCLEON'S FP - I.I. - P.cl. COMP. CONSTRUCTION BY MICROWAVE

PATHOLOGY OF MICROWAVE –LASER INJURIES

POWERFUL DESTRUCTIVE BIOCIDAL SUPERSONIC MICROWAVE FUNDAMENTAL PARTICLES INTER INTO INSIDE CNS AUDITARY CENTERS INTERNAL ELECTRONS- NUCLEONS SUBSYSTEM UNITS CHEMICAL LAB.S AND DISRUPT FINE S. Y. – F.P. – I.I. – P.cl. _ COMPOUNDS CONSTRUCTIONS IN CNS ELECTRONS -NUCLEONS, MOST OF THESE PARTICLE-CLOUD COMPOUND CONSTRUCTIONS ARE IMPORTANT FACTORS AT PRODUCTION OF THOUGHT CURRENTS WHICH ARE S. Y. – F.P. I.I. – P.cl. CURRENTS OF THE DIFFERENT KINDS CONSTRUCTED THROUGH DIFFERENT NORMAL Y. S.- F.P.- I.I.- P.cl. –COMPOUND COSTRUCTIONS HISTORY OF MAN KIND DURING PAST HUMAN LIFE HISTORY,

INTERANCE OF DESTRUCTIVE BIOHOSTILE POWERFUL MICROWAVE FUNDAMENTAL PARTICLE INTO CNS-ELECTRONS AND NUCLEONS UPON INTERANCE TEAR APART AND BREAK DOWN MOST OF FINELY PRE-CONSTRUCTED BIOFRIENDLY PRE-EXISTING PREVIOUSLY CONSTRUCTED Y. S. –F.P. –I.I. – P.cl. _ COMPOUND CONSTRUCTIONS OF CNS ELECTRONS - NUCLEONS,

IN CONTINUATION OF CNS ELECTRON-NUCLEON PARTICLE COMPOUND DESTRUCTIONS, THE HARSH BIO-CIDAL MICROWAVE PARTICLES COMBINE WITH REMNANTS OF INTER ELECTRON- NUCLEON PARTICLE COMPOUNDS UNDER REGENERATIVE OR DEGENERATIVE A. S. I. F.P. Mol. C.I.C. AND PRODUCE NEWLY CONSTRUCTED NON-LIVE MICROWAVE PARTICLE-COMPOUNDS CONSTRUCTIONS INSIDE CNS AUDITARY CENTERS ELECTRON-NUCLEONS,

THE PRODUCED ROARING MICROWAVE PARTICLE COMPOUND CONSTRUCTIONS INSIDE CNS ELECTRONS – NUCLEONS CAUSE CREATIONS OF PERSISTENT HARSH ROARING TINITUS IN PATIENTS HEAD WHICH THESE TINITUS ARE NON INTERRUPTIVE AND PERMANET TINITUS WITH NO RELIEF, AND RELATE TO EXISTENCES OF MICROWAVE PARTICLES COMPOUNDS IN CNS ELECTRONS-NUCLEONS PARTICLE COMPOUND CONSTRUCTIONS,

THE PHENOMENON OF NON-LIVE BIOCIDAL MICROWAVE PARTICLE-COMPOUND NEO-GENESIS AND CONSTRUCTIONS OF CNS- ELECTRONS AND NUCLEONS PARTICLE COMPOUND WITH MICROWAVE PARTICLE CONSTRUCTIONS ARE MAIN CAUSE OF PRODUCTION OF PERSISTING HARSH NON STOP TINITUS SECONDARY TO EXISTING MICROWAVE PARTICLE COMPOUNDS INSIDE BRAIN ATOMS WHICH PRODUCES NON-STOP TINITUS INSIDE HEAD WITH NO RELIEF,

PHYSIOLOGICAL FUNCTIONS OF CERUMEN

WHEN PATIENT'S HEAD SHOT BY WEAPON GRADE MICROWAVE FROM ANY DISTANCES, THE MICROWAVE -LASER PARTICLES INTER INTO HEADTHROUGH EXTERNAL EAR CANALS, AND THROUGH EXTERNAL EAR CANAL, MIDDLE EAR, AND INNER EAR TRANSIT DIRECT INTO CNS AUDITARY CENTER INTERNAL ELECTRON-NUCLEON STRUCTURES, POWERFUL MICROWAVE DESTRUCT EVERY STRUCTURE IN THE PATH, DEPENDING WHAT KIND AND WHAT DOSE, MICROWAVE SHOT INTO HEAD, AND HOW MUCH AND HOW LONG CONTINUOUSLY RUNNING INTO BRAIN.

AT EXTERNAL EAR CANAL THE COPIOUS EXCRETION OF THE CERUMEN IS THE MOST IMPORTANT DEFENSE SYSTEMS TO DIMINISH THE BIOHOSTILE SONIC PARTICLES DESTRUCTIONS, CERUMEN'S MAIN JOB IS NOT THE INSECT REPELLENT. IN ORDER TO DIMINSH HARSH POWERFUL DESTRUCTIVE VIBRATIONS AND ROARING POWERFUL CURRENTS OF MICROWAVE PARTICLE LASER CURRENTS, AND PROTECT MIDDLE EARS FINE BONY STRUCTURES FROM DISRUPTIONS ,AND TYMPANIC MEMBRANE RUPTURE, INJURIES OF CNS AND INNER EAR NEURAL CONSTRUCTIONS IS COPIOUS CERUMEN EXCRETION BY EXTERNAL EAR, WHICH COATS ALL AROUND THE EXTERNAL EAR CANAL WITH INSTANT EXCRETIONS AND COVER T.M. AS WELL, AND HELP TO REDUCTIONS OF DESTRUCTIVE

POWERS OF THE MICROWAVE.THIS IS THE MAIN FUNCTION AND ESSENTIAL PHYSIOLOGICAL FUNCTIONS OF EAR SERUMEN EXCRETIONS.

THE EXTERNAL EAR GLANDS EXCRETE COPIOUS QUANTITIES OF CERUMEN AND COVER THE T. M. AND EXTERNAL EAR CANANL LUMEN ALL AROUND , AND DIMINISH ROARING HARSH DESTRUCTIVE SHEARING FREQUENCIES, AND COMBAT BIOHOSTILE MICROWAVE VICIOUS SONIC – FUNDAMENTAL PARTICLES WHICH TEAR APART ANY FINE BIOLOGICAL TISSUES IN ITS PATH, THIS PHENOMENON OF CERUMEN EXCRETIONS AND EXISTENCE IN EAR CANAL IS CREATED TO COMBAT AND CONTROL BIOHOSTILE HARSH SONIC PARTICLE INJURIES AND PROTECT SENSITIVE VITAL STRUCTUTRES OF CNS AUDITARY SYSTEMS,

THE EXCRETED CERUMEN COVER LUMENS OF EXTERNAL EAR CANAL AND TYMPANIC MEMBRANE, TO DIMINISH DESTRUCTIVE FORCES OF BIOHOSTILE INDUSTRIAL MICROWAVE PARTICLE CURRENTS, TO PREVENT SHEARING ACTIONS OF HARSH VIBRATING FREQUENCIES OF MICROWAVE PARTICLES IN ORDER THE T.M., FINE EARS BONY STRUCTURES AND FINE NEURAL TISSUES OF INNER EAR AND BRAIN NOT TORN APART INTO PIECES BY INDUSTRIAL MICROWAVE FORCES, THE CERUMENS MAIN BIOLOGICAL FUNCTIONS IS COMBATING AGAINST BIOHOSTILE SONIC AND MICROWAVE PARTICLES IN ORDER PREVENT THEIR DESTRUCTIONS OF AUDITARY CONSTRUCTIONS,

THE WAR SYNDROMES

AND WAR CRIMES AGAINST HUMANITY

SYMPTOMS AND SIGNS OF MICROWAVE WEAPONS BRAIN-INJURIES

SYMPTOMS AND SIGNS OF PATIENTS WHO HAS BEEN SHOTS BY MICROWAVE WEAPONS OF MASS DESTRUCTION ON HEAD, AND BRAIN INJURIES, CHRONIC DESTRUCTION OF CNS ELECTRONS NUCLEONS:

WHEN DURATION OF BIOHOSTILE POWERFUL MICROWAVE SHOTS ON HEAD CONTINUES FOR HOURS OR MORE, PATIENT BECOME RESTLESS, THE POTENT HIGH POWER INDUSTRIAL BIOCIDAL LASER- MICROWAVE SHOTS ON HEAD OF INDIVIDUALS WITH BORDER LINE HEALTH CONDITIONS PRODUCE CONTINUOUS STIFF NECK, NAUCEA, VOMITING HEADACHES SIMILAR TO MENINGEAL IRRITATION SIGNS AND SYMPTOMS, BUT IN PARA-CLINICAL TESTS THE FINDINGS ARE NON- CONCLUSIVE AND NORMAL LOOKING, MICROWAVE DAMAGES ARE INTERNAL ELECTRON-NUCLEON INJURIES BY OBNOXIOUS BIOHOSTILE PARTICLES BUT MENINGITIS PRODUCED BY MICRO-ORGANISM OR OTHER SIMILAR CAUSES WITH CELLS AND TISSUE DAMAGE INVOLVEMENTS.

THE INDIVIDUALS WHO ARE YOUNG, STRONG, HEALTHY WITH GOOD BODY IMMUNITY, RESIST BEING KILLED BY MICROWAVE INSULTS.

BUT MOST WITHOUT EXCEPTIONS SUFFER FROM VARIABLE DIFFERENT MICROWAVE CHRONIC COMPLICATIONS, AND PERMANENT DISABILITIES SUCH AS NERVE DEAFNESS, PERMANENT NON - STOP TINITUS, AND CHRONIC DIFFERENT DEGREES BRAIN DAMAGES.

THE SYMPTOMS AND COMPLAINTS THAT CAUSED BY MICROWAVE INJURIES, ARE SIMILAR TO THE SYMPTOMS AND COMPLAINTS WHICH PRODUCED THROUGH THE OTHER DISEASES OF BRAIN DAMGE, LIKE: MENINGEAL INFECTIONS TYPE COMPLAINTS, GUN SHOT PRODUCED SYMPTOMS, BLUNT INJURIES TYPE COMPLAINT, ETC.).

GENERALLY, MICROWAVE DOE'S NOT LEAVE DEMONSTRABLE SIGNS TO PROVE THE EVIDENCE OF MICROWAVE INJURIES IN BEHIND, NO CUTS SIGNS, NO MASS OR REDNESS SIGNS, ETC. THE INJURED BRAIN BY MICROWAVE NEED PROLOGED PROPHYLAXIS ANTIBIOTIC TRAETMENTS, BECAUSE INJURED TISSUE BY MICROVAWE IS SIMILAR TO

RADIATION WOUND, DO NOT HEAL FAST, MICROWAVE DAMAGES MOSTLY ARE PRONE TO SECONDARY INFECTIONS ETC. ALSO MICROWAVE INJURIES HEALS VERY SLOWLY OVER YEARS TO RECONSTRUCT ELECTRONS NUCLEONS CONSTRUCTIONS, ALSO THE HEALING PROCESS MAY NEVER OCCUR AS WELL AFTER LONG TIMES IN SOME CASES.

WHEN PATIENTS SHOT BY MICROWAVE, ALWAYS PATIENTS FEEL TRANSITION PATHS OF HARSH POWERFUL ROAMING MICROWAVE CURRENTS, PASSING THROUGH EXTERNAL EAR CANAL TOWARD THE AUDITARY CNS CENTERS, AND THE PATIENTS CAN SHOW WITH FINGER POINT THE PLACE OF PAIN OVER SKULL, WHICH IT CORRELATES TO CNS HEARING CENTER LOCATION AT BRAIN.

THE MICROWAVE DEAFNESS AND HARSH CONTINUOUS TINITUS ROARS WITH NO STOP ALWAYS IS THE FIRST COMPLAINT THE PATIENT STATES WHEN SHOT BY MICROWAVE ON HEAD. PATIENTS MOSTLY IN SERIOUS INJURIES MANIFEST BRAIN INJURY SIGNS AND SYMPTOMS.

THE SYMPTONS IN SERIOUS CASES ARE HEADACHE, STIFF NECK, NAUSIA, VOMITING, CONFUSION, FEVER, REMARKABLE NECK RIGIDITY, GRADUAL HEARING LOSS, CONTINUOUS HARSH ROARING WITH NO STOP TWENTY - FOUR HOURS, ETC. BUT CHANGES IN XRAYS, LAB. TESTS NON SPECIFIC, MAY NOT HELP. THE MRI MAY NOT HELP TO DIAGNOSIS.

FACTS ABOUT THE INTERNATIONAL MICROWAVE PARTICLE WARS AND WAR SYNDROMES

IN MODERN INTERNATIONAL WARS, SHOOTING WITH MICROWAVE FUNDAMENTAL PARTICLE'S CODED WITH LIGHT – LASER WEAPONS (BULLETS). OR UNSING ANY OTHER BIOHOSTILE PARTICLES, THESE WEAPONS CONTAMINATE AIR AND SPACE WITH OBNOXIOUS PARTICLE CLOUDS. AND THESE ARE WEAPONS OF MASS DESTRUCTIONS AND CRIME AGAINST HUMANITY THAT DAMAGES ORDINARY PEOPLES, CHILDREN, WOMEN, MEN ALIKE AS THE SOLDIERS ARE NOT AWARE ABOUT THE PARTICLES IN AIR, THAT THEY ARE LIVING AND BREATHING.

WHEN AIRBORN MICROWAVE- LASER PARTICLES INTER INTO AUDITARY CNS CENTERS INTERNAL ELECTRON-NUCLEON, DESTRUCT AND TEAR DOWN NORMAL S.Y. – F.P. – I.I. - PARTICLE CLOUD –COMPOUND CONSTRUCTION INSIDE BRAINS ELECTRONS –NUCLEONS. THE CNS INTERNAL ELECTRON-NUCLEON S. Y. – F.P. – I.I.- P. cl. _ COMPOUND DESTRUCTIONS.THE SYMPTOMS AND SIGNS ARE SIMILAR TO THE THOSE SYMPTOMS AND SIGNS OF NEUROLOGICAL BRAIN INJURY DISEASES.

THE SOLDIERS AND NATIVE CITIZENS MAY SUFFERE FROM ACUTE PARTICLE INJURIES SYNDROMES WHEN THEY ARE IN NEIBORHOOD OR IN FIELD, OR THEY MAY DEVELOPE CHRONIC COMPLICATIONS IN LATER TIMES.

The Microwave PARTICLE CURRENT'S TRAVEL IN TISSUES

PARTICLE SHOTS INTO CHEST CAVITY AND INTER- ABDOMINAL ORGANS

PATTERNS OF MICROWAVE LASER FUNDAMENTAL PARTICLES TRANSIT INSIDE BODY TISSUES:

MICROWAVE PARTICLE'S TRAVEL THROUGH TISSUES:

INSIDE TISSUES FUNDAMENTAL PARTICLES CURRENTS FLOW, IN STAIGHT LINES, AS SEPARATE MULTIPLE PARALLEL FUNDAMENTAL PARTICLE CURRENTS, IN FORMS OF MULTI -PARALLEL PARTICLE CURRENT FLOWS SEPARATE AND APART FROM EACH OTHER. AT STRAIGHT LINES, BETWEEN THE PARTICLE'S ENTERANCES POINTS AND EXIT POINTS.

INSIDE CHEST CAVITY BIOHOSTILE POWERFUL MICROWAVE PARTICLES DO NOT FLOW AS ONE SINGLE PARTICLE CURRENT FROM ONE SIDE TO OPPOSITE POINT, THE POWERFUL OBNOXIOUS MICROWAVE PARTICLES FLOW INSIDE THE CHEST AND ABDOMINAL CAVITY AS SEPARATE MULTI – PARALLEL PARTICLE CURRENTS APART FROM EACH OTHER, IN MULTI -STRAIGHT LINE SEPARATE INDEPENDENT FLOWS.

BENIGN ULTRASONIC PARTICLE CURRENTS ARE NON SENSIBLE. IN CONTRAST THE BIOHOSTILE MICROWAVE – LASERFUNDAMENTAL PARTICLE CURRENTS ARE SENSIBLE. THE ALL POWERFULL BIOHOSTILE MICROWAVE PARTICLE CURRENTS, AS WELL AS BIOHOSTILE ELECTRIC CURRENTS ALL ARE SNESIBLE, WHEN TRAVELING INSIDE BODY TISSUES. MICROWAVE LASER CURRENTS ALL ARE SENSIBLE WHEN CROSSING INSIDE THE HEART, LUNG, LIVER, AND OTHER BODY ORGANS.

THIS SENSIBILITIES OF INDUSTRIAL POWER BIOHOSTILE MICROWAVE PARTICLES, THROUGH SENSING THE PARTICLES REVEALS, THAT THE BIOHOSTILE MICROWAVE PARTICLES CURRENTS ARE NOT SINGLE ONE STREAM PARTICLE FLOWS INSIDE BODY TISSUES, BUT BIOHOSTILE MICROWAVE PARTICLES CURRENTS FLOW THROUGH: MULTIPLE DIFFERENT PARALLEL CURRENTS ONE NEXT TO OTHER IN STRAIGHT LINES BETWEEN INTERANCE POINT TO CHEST, BELLY AND BODY, AND STAY SENSIBLE THROUGH OUT ALL INTERRNAL BODY TRANSIT TIME, ALL INTERNAL BODY TRAVEL PATH, UNTIL IT EXIT FROM OTHERSIDE OF BODY THROUGH EXIT POINT, ALL ARE SENSIBLE.

THE HUMAN SPECIES WILL SENSE PAINLESSLY THE BIOHOSTILE PARTICLES, WHEN THESE MICROWAVE PARTICLE CURRENTS SHOTS INTO HUMAN'S CHEST AND ABDOMEN AND PARTICLE CURRENTS TRAVELING INSIDE DIFFERENT THORACO-ABDOMINAL TISSUES ALL ARE CLEARLY SENSIBLE.

WHEN THE BIOHOSTILE MICROWAVE PARTICLE CURRENTS INTER INTO THE CHEST CAVITY, AND FLOWING AS SEPARATE MULTIPLE PARALLEL PARTICLE CURRENT FLOWS, IN STRAIGHT LINES FROM POINT OF INTERANCE TO THE CHEST WALL DIRECT TO THE EXIT POINT OF BIOHOSTILE PARTICLES EXIT POINT FROM OTHERSIDE TO OUTSIDE BODY.

THE HUMAN BODY WILL SENSE PAINLESSLY THE FLOW OF BIOHOSTILE PARTICLE CURRRENTS INSIDE DIFFERENT INTERNAL BODY ORGANS SUCH AS: HEART, LUNGS, LIVER, AND INSIDE ABDOMINAL ORGANS WHEN THE MICROWAVE CURRENTS FLOWING AND CROSSING DIFFERENT BODY TISSUES. FINALLY EXITING FROM THE OPPOSITE SIDES OF ENTERANCES TO OUTSIDE.

BIOHOSTILE MICROWAVE PARTICLES CURRENTS INTERACTIONS WITH BODY ORGANS – PCS:

THE POWERFUL PARALLEL, SEPARATE PARTICLE CURRENTS OF MICROWAVE SUPERSONIC FUNDAMENTAL PARTICLES INSIDE THE CHEST CAVITY ALL FLOW IN STRAIGHT LINES, AS SEPARATE CURRENTS PARALLEL TO EACH OTHER, BETWEEN THE ENTERANCE POINT OF THORACIC WALL IN ONE SIDE OF THORAX, DIRECT LINE FLOW TO THE EXIST POINT IN THE OPPOSITE SIDE OF THE THORACIC CAGE.

THE MICROWAVE PARTICLE FLOW INSIDE THE THORACIC CAVITY TISSUES ARE MULTIPLE PARALLEL, SEPARATE FLOWS, ALL DIFFERENT PARALLEL CURRENTS CROSS AND INTER INTO DIFFERENT INTER –THORACIC OR INTER-ABDOMINAL TISSUE IN STRAIGHT LINES FROM ONE POINT INTERANCE TO OPPOSITE SIDE OF EXIT, THESE DIFFERENT PARALLEL STRAIGHT LINE PARTICLE FLOWS DO NOT JOIN TOGETHER, AND DO NOT FLOW AS ONE LARGE SINGLE STREAM FLOW, OR SINGLE ONE LARGE CURRENT IN THORACIC CAVITY TISSUES. THIS PATTERN OF PARTICLE CURRENTS ALL SENSED CLEARLY BY VICTIMS. THESE ARE NUMEROUSLY SENSED FACTS AT OVER HUNDREDS EPISODES.

AS BEFORE EXPLAINED, THE ALL ANIMALS, LIVING THINGS ORGAN'S FUNCTION OPERATES UNDER THEIR ORGAN-SPECIFIC PARTICLE INTELLIGENCE SYSTEM CENTERS, AND ALL ELECTRONS NUCLEONS POPULATIONS OF THE ORGAN IS CONNECTED TO EACH OTHER, THROUGH ORGAN SPECIFIC PARTICLE CIRCULATION SYSTEMS.

WHEN THE POWERFUL LETHAL DOSES BIOHOSTILE LIGHT LASER CODED MICROWAVE SHOTS INTERING INTO DIFFERENT THORACO-ABDOMINAL ORGANS AND SYSTEMS, DEFFINITELY THESE BIOHOSTILE PARTICLE CURRENTS WILL INTERFERE WITH THESE VITAL ORGANS PARTICLE CIRCULATION SYSTEMS (PCS), SPECIFICLY IN ELDERLY, AND BORDERLINE HEALTH CONDITION INDIVIDUALS.

IN ELDERLY AND BORDER LINE HEALTH CONDITIONS, THE BIOHOSTILE MICROWAVE PARTICLES PRODUCE SERIOUS IRREVERSIBLE MEDICAL COMPLICATIONS SUCH AS DYSRHYTHMIA, WHEN BIOHOSTILE LASER MICROWAVE PARTICLES INTER INTO HEART TISSUES AND CROSS CARDIAC ORGAN, CAN DISRRUPT CARDIAC ELECTRICAL PARTICLE CIRCULATION SYSTEMS OF THE HESS- INTELLIGENCE SYSTEM CENTER, THE HESS INTELLIGENCE SYSTEM CENTER THROUGH CARDIAC PARTICLE CIRCULATION SYSTEMS (P.C.S.) CONTROLS ALL INTER -CARDIAC ELECTRONS NUCLEONS TOTAL POPULATIONS NANO – FUNCTIONS. THE POWERFUL STRONG BIOHOSTILE MICROWAVE LASER PARTICLES DISRUPT CARDIAC PARTICLES CIRCULATIONS SYSTEMS FROM HESS CENTRAL INTELLIGENCE SYSTEM CENTERS, OR FROM ANY OTHER MAJOR BRANCHES TO ELECTRICAL FUNDAMENTAL PARTICLE CIRCULATION SYSTEMS OF THE HEART, ANY DISRUPTION OF HESS – P. C. S. CAN CAUSE CARDIAC FUNCTION DISORDERS AND COMPLICATIONS SUCH AS: DYSRHYTHMIA, C. H. F., M. I., CARDIAC ARREST, AND DEATHS.

THE PHENOMENONS SIMILAR TO HEART MAY OCCUR, IN OTHER INTRA-THORACIC ORGANS, IN LUNGS, AND OTHERS, UNDER LETHAL DOSES OF BIOHOSTILE MICROWAVE PARTICLES SHOTS. MICROWAVE EASILY CAN TEAR DOWN WHOLE OPERATTIONS OF DIFFERENT BODY ORGANS AND SYSTEMS PARTICLE CIRCULATIONS SYSTEMS TOTAL OPERATIONS IN SECONDS, AND CAUSE ALL ORAGANS AND SYSTEMS TOTAL ELECTRONS NUCLEONS POPULATIONS NANO-FUNCTIONS DISAPPEAR, THAT MEANS NO BIOLOGICAL EXISTANCE AND FUNCTIONS AND DEATH.

THE WEAPONS WHOSE INJURIES AND MURDERS DO'S NOT LEAVE A TRACE FROM THE COMMINTED CRIMES!

IN THE CASES OF BIOHOSTILE STRONG LASER - MICROWAVE SHOT - INJURIES ON HEAD WHEN SHOOTING DIRECT ON THE HEAD CAUSE IRREPAIRABLE DAMAGES FREQUENTLY PRODUCE HEADACHES, STIFF NECK, NAUSIA LOCAL TENDERNESS EXACTLY ON SKULL SIDE OVER THE INJURED CNS- ELECTRONS- NUCLEONS INJURED TISSUES AND LOW LEVEL FEVER ETC. ALL ARE DETECTABLE BY EXAMINERS,

IN CONTRAST THE CHEST INJURIES DOES NOT PRODUCE TOO MUCH SIGNS OR SYMPTOMS IN NORMAL STRONG INDIVIDUALS, UNLESS THE OCCURANCES OF COMPLICATIONS SUCH AS SUDDEN MI, CHF, CARDIO-PULMONARY ARREST OR SIMILAR SITUATIONS AND DEATH WHEN OCCUR SPECIFICLY IN BORDERLINE HEALTH INDIVIDUALS WHEN HIT BY POWERFUL INDUSTRIAL SUPRASONIC PARTICLE CURRENTS ON HEART AND VITAL ORGANS THROUGH PRODUCTION OF DYSRHYTHMIAS, BLOCKING CARDIAC RHYTHM CONDUCTIONS AND PRODUCING HEART –ARREST SECONDARILY CAUSING M.I., ETC., MOSTLY OCCUR IN BORDERLINE HEALTH INDIVIDUALS, MOSTLY PRONOUNCED DEAD BY NATURAL CAUSES AND CERTIFICATES ISSUED ROUTINELY. THIS SITUATION MAKES DESPOTS TO MURDER AND TORTURE FREELY WITHOUT NO RESTRICTIONS FREELY IN THIRD WORLD OR SOCIALIST NATIONS.

MICROWAVE SHOT OF BORDERLINE INDIVIDUALS EASILY CAN PRODUCE AND PROVOKE DYSRHYTHMIC CARDIOPULMONARY MALFUNCTIONS AND DEATH, CARDIAC ARREST, RHYTHM FAILURE, AND SECONDARILY ALL OF ABOVE PRODUCE DETECTABLE DISEASES SUCH AS MI, OTHER CARDIOPULMONARY INCIDENTS, WHICH THERE IS NO DIFFERENCES BETWEEN THESE LASER MICROWAVE DEATHS AND NATURALLY OCCURING DEATHS. ALL LOOK SIMILAR, THIS PHENOMENON ARE ABUSED EXTENSIVELY BY SOME AUTHORITIES AGAINST PEOPLE. IN SOCIALIST BACKED NATIONS.

DESPOTS ABUSE MICROWAVE-LASER WEAPONS OF MASS DESTRUCTION AGAINST OWN PEOPLE, WHO MAY HAVE DIFFERENT IDEAS THAN THEM. UNDER THESE DIFFERENCES THE RULERS IN ORDER TO RULE CRIMES, THEY KILL INTELLECTUALS, THROUGH USE OF MICROWAVE-LASER WEAPONS, WHICH MICROWAVE SHOTS, DOES NOT LEAVE A TRACE IN BEHIND FROM THESE MURDERS.

THERAPY OF MICROWAVE INJURIES

===

PATHOLOGY OF CNS ELECTRONS – NUCLEONS DAMAGES BY MICROWAVE

THE MAIN PATHOLOGICAL DAMAGES OF WEAPONIZED MICROWAVE INJURIES INTO AUDITARY CNS ELECTONS-NUCLEONS PARTICLE COMPOUNDS CONSTRUCTIONS, AND OTHER CNS CENTERS, MOSTLY IT IS IN DISRUPTION AND DESTRUCTIONS OF THE CNS S. F.P. – I.I. – P. cl. _ COMPOUND MOLECULAR CONSTRUCTIONS INSIDE CNS ELECTRONS AND NUCLEONS, SPECIALLY INSIDE THE AUDITARY CNS ELECTRONS AND NUCLEONS.

THE POWERFUL BIOHOSTILE MICROWAVE PARTICLES TEAR DOWN NORMAL S. –F.P. – I.I. – P. cl. – COMPOUNDS OF CNS AUDITARY ELECTRONS NUCLEONS COMPOUNDS. AND IN THEIR PLACE, STORAGE NON-LIVE BIOHOSTILE MICROWAVE PARTICLE COMPOUNDS CONSTRUCT HARMFUL PATOLOGICAL MOLECULAR CONSTRUCTIONS. WHICH CAUSE HARSH CONTINUOUS ROAR WITH NO INTRUPTIONS IS THE RESULTS.

THE NORMAL ELECTRON NUCLEON S. –F.P. – I.I. – P. cl. COMPOUND CONSTRUCTIONS OF AUDITARY CNS CENTER, FORCEFULLY TORN DOWN, DISABLED AND REMOVED BY POWERFUL MICROWAVE LASER PARTICLES. THEREAFTER THE NORMAL STRUCTURES HAVE BEEN DESTRUCTED AND REPLACED BY NON-LIVE IMPLANTED DEAD NON - FUNCTIONAL ROARING MICROWAVE PARTICLE-COMPOUNDS, AND COMPOSITES. WHICH THEIR COMPLICATION IS UNABLE TO HEAR AND UNDERSTAND, AND NON FUNCTIONAL BRAIN PARTIALLY, AS WELL AS NON- INTERRUPTED CONTINUOUS TINNITUS WITH NO STOP

THE THERAPY FOR THIS PROBLEM IS ELIMINATION AND REMOVING ROARING DESTRUCTIVE MICROWAVE PARTICLE-COMPOUNDS OUT OF AUDITARY CNS INTER CNS ELECTRONS - NUCLEONS CONSTRUCTIONS, THROUGH DEGENERATIVE A. S. I. F.P. Mol. C.I.C.

AFTER REMOVING OF THE ABNORMAL PARTICLE CLOUDS, IN THEIR PLACE MUST CONSTRUCT AND IMPLANT NORMAL S. –F.P. – I.I. – P. cl. _COMPOUNDS THROUGH REGENERATIVE A. S. I. F.P. – Mol. C.I.C. INSIDE THE AUDITARY CNS INTERNAL ELECTRONS- NUCLEON PARTICLE COMPOUND, THIS IS RE- CONSTRUCTION OF CNS ELECTRONS NUCLEONS PARTICLE AGAIN, RESTORING THE SONIC PARTICLE CLOUDS STORAGE SYSTEMS THROUGH USES AND STORAGES OF NORMAL SONIC PARTICLE CLOUDS. THROUGH THE USES OF NORMAL PARTICLE CLOUDS TO BUILD NORMAL S. – F.P. – I.I.- P. cl. COMPOUNDS INSIDE THE CNS AUDITARY CENTER'S ELECTRONS AND NUCLEONS.

THIS PROCEDURES WILL REVERCE PREVIOUSLY PRODUCED NEURAL DEAFNESS, AS WELL AS WITH REMOVAL OF HARSH MICROWAVE FROM CNS ELECTRONS NUCLEONS COMPOUNDS, THE HARSH CONTINUOUS TINNITUS ALSO GRADUALLY ELEIMINATE AND DIMINISH. ABOVE IS TREATMENT OF CHOICE FOR MICROWAVE INJURIES OF CNS INTERNAL ELECTRON NUCLEON REPAIRES.

EZZAT E. MAJD POUR, M.D.-

ADDRESS: ezzatmajd@icloud.com

DISCOVERIES OF AUTHOR DURING COLLEGE YEARS:

1 - I DISCOVERED FUNDAMENTAL PARTICLES (FP) CONSTRUCT ELECTRONS, NUCLEONS, ATOMS, AND NANO UNITS UNDER A. S. I. FP. Mol. CIC. I DISCOVERED INTERNAL ELECTRONS, NUCLEONS, ATOM'S PARTICLE INTELLIGENCE SYSTEM CENTERS (PIS), ALSO INTERNAL ELECTRONS, NEUTRONS, PROTON'S PARTICLE CIRCULATION SYSTEMS (PCS), ELECTRONS NUCLEON'S PSYCHE GENESIS, AND FACTUAL OR REAL INTELLIGENCE (RI).

2 - I DISCOVERED FUNDAMENTAL PARTICLE'S GENERAL CHEMISTRY, ORGANIC CHEMISTRY, AND BIOLOGICAL CHEMISTRY. ALSO HOW EXOGENOUS AIRBORNE PARTICLE CLOUDS ENTER INTO ELECTRONS, NUCLEONS SUBSYSTEM- UNITS CHEMICAL LAB.S, COMBINE WITH INTERNAL ELECTRON, NUCLEON PARTICLE COMPOUNDS, AND CONSTRUCT MOLECULAR COMPOUND CONSTRUCTIONS OF PROTONS, NEUTRONS, AND ELECTRONS. AND THIS IS PHENOMENON OF ELECTRON GENESIS, PROTON GENESIS, NEUTRON GENESIS.

3 – I DISCOVERED MANY PLANTS, ALSO POSSESSES SENSARY INTELLIGENCE SYSTEMS, MOTOR INTELLIGENCE SYSTEMS, AND PLANTS CENTRAL INTELLIGENCE SYSTEMS. ALSO CHEMICAL COMBINATIONS OF THE AIRBORNE LIGHT FUNDAMENTAL PARTICLES WITH INTERNAL CELL ELECTRONS, NUCLEONS MOLECULAR COMPOUNDS, UNDER A. S. I. FP. Mol. CIC, PRODUCE LIGHT PARTICLE COMPOUNDS CONSTRUCTIONS OF THE PLANTS ATOMS, ELECTRONS, NUCLEONS. THE PHENOMENON OF PLANTS ATOM GENESIS, AND PLANTS CELL GENESIS.

4- DURING STUDY IN PHYSICS, I DISCOVERED EXOGENOUS AIRBORNE AND INDIGENOUS PARTICLE CLOUD CIRCULATION SYSTEMS, TRANSMISSION OF THE PARTICLE CLOUDS AND FUNDAMENTAL PARTICLES THROUGH PARTICLE CURRENTS BETWEEN THE DIFFERENT LIVE AND NON-LIVE PLANETARY ELECTRONS, NEUTRONS, PROTONS, AND ATOMS, COMBINE WITH INTERNAL ELECTRONS-NUCLEON PARTICLE COMPOUNDS, AND CONSTRUCT ELECTRONS, NUCLEONS MOLECULAR STRUCTURES (PHENOMENON OF ELECTRON-GENESIS, NUCLEON –GENESIS).

5 – I DISCOVERED MOLECULAR EVOLUTION FIRST STARTED BETWEEN FUNDAMENTAL PARTICLES, UNDER A. S. I. FP. Mol. CIC. CONSTRUCTED LARGER MOLECULAR NANO-UNIT CONSTRUCTIONS, IN SECOND PHASE EVOLUTION WAS BETWEEN ATOM BASE MOLECULAR STRUCTURES, AND AT THIRD TO Nth. PHASES OF MOLECULAR EVOLUTION TOOK PLACE BETWEEN BIOMOLECULAR AND MICRO- MOLECULAR EVOLUTION STRUCTURES, AND CONTINUED UP TO MACRO- MASSES EVOLUTION PHASES, PRODUCED EXISTING PLANETARY DIFFERENT SUBJECTS.

6 – I DISCOVERED EMBRYO GENESIS PROCESS IN UTERUS ARE EQUAL TO THE GENESIS OF LIVING THINGS WHICH TOOK PLACE DURING PAST WORLD HISTORY UNDER THE MOLECULAR EVOLUTION ERAS. UNDER MOLECULAR EVOLUTION ORDERS AND PATHS AND CREATED EXISTING LIVING THINGS.

7 – I DISCOVERED ALTERNATING AUTONOMOUS SEQUENTIAL STABLE STEM CELLS MUTATIONS IN EMBRYO PRODUCE ALTERNATING CELLS MEDIA CHEMICAL FORMULARY CHANGES, THE EMBRYONIC MEDIA CHEMICAL FORMULARY CHANGES SECONDARILY PRODUCE CELL MUTATION. AND CAUSE STEM CELL DIFFERENTIATIONS, AND IS THE MAJOR CAUSE FOR EMBRYO-GENESIS, EQUAL TO THE MOLECULAR EVOLUTION ORDERS.

DISCOVERIES OF THE AUTHOR DURING THE MEDICAL SCHOOL YEARS:

I DISCOVERED HOW CHEMICAL COMBINATIONS OF EXOGENOUS S. Y. T. E.- FP – I.I. – P. Cl. WITH CNS INTERNAL ELECTRON NUCLEON PARTICLE COMPOUNDS UNDER A. S. I. FP. Mol. CIC. STORE INTELLIGENCE AND KNOWLEDGE INSIDE THE CNS ELECTRONS AND NUCLEONS IN PARTICLE CLOUD COMPOUND FORMS, AND HOW INTERACTIONS OF THESE PARTICLE CLOUDS BETWEEN DIFFERENT CNS CENTERS ELECTRONS AND NUCLEONS PRODUCE THOUGHT CURRENTS AND PSYCHE (PSYCHE-GENESIS PHENOMENON). ALSO I DISCOVERED PSYCHIATRIC DISEASES CAUSED BY PARTICLE CLOUD TRANSMISSION FROM ONE PERSON TO OTHER, AND ARE CONTAGIOUS.

AFTER GRADUATION FROM SCHOOL, I DISCOVERED HUNDREDS OF OTHER DISCOVERIES, WHICH NO ONE HAS DONE THOSE DISCOVERIES BEFORE, SEE DIFFERENT AUTHORS BOOKS IN DIFFERENT SUBJECTS.

EZZAT E. MAJD POUR, M.D.